THE UNIVERSITY OF WINNIPEG

LIBRARY FUND

IN MEMORY
OF

Cecil M. Fung

Presented by

Delza Longman

Modern European History

A Garland Series of Outstanding Dissertations

MODERN EUROPEAN HISTORY

A Technological Lag

Diffusion of Electrical Technology in England, 1879–1914

A.J. Millard

Garland Publishing, Inc.
New York and London 1987

Library of Congress Cataloging-in-Publication Data

Millard, A. J.
A technological lag : diffusion of electrical technology in England, 1879–1914 / A.J. Millard.

p. cm.—(Modern European history)
Based on the author's thesis (Ph. D.)—Emory University, 1981, presented under title: The diffusion of electric power technology in England, 1880–1914.
Bibliography: p.
Includes index.
ISBN 0-8240-7823-3 (alk. paper)
1. Electric engineering—England—History.
I. Title. II. Series.
TK18.M55 1987 87-25195 CIP
942′.081—dc 19

All volumes in this series are printed on acid-free, 250-year-life paper.

Printed in the United States of America

A TECHNOLOGICAL LAG

DIFFUSION OF ELECTRICAL TECHNOLOGY IN ENGLAND 1879-1914

by

A.J. Millard

CONTENTS

PREFACE

This book began as my dissertation, "The Diffusion of Electric Power Technology in England, 1880-1914," which was submitted to Emory University in 1981. Rondo Cameron, William Rand Kenan University Professor, directed the dissertation and guided me through all stages of research and writing.

A grant from the National Science Foundation supported some of my research and I benefited from a post doctoral grant from the National Endowment for the Humanities.

The academic support services of Bentley College provided me with the means to turn the dissertation into this manuscript. I owe a great debt to Anna Cibotti who typed the manuscript. Judy Edwards provided invaluable technical assistance for the word processor and laser printer.

Andre Millard
Waltham, Mass.

CHAPTER ONE

THE LITERATURE OF TECHNOLOGICAL LAGS

The process of industrialization has been carried forward by the growth of "leading sectors" in the economy. According to Rostow, the classic sequence of great leading sectors has been cotton textiles, railroads, iron and steel, chemicals and electricity, and the automobile industry.[1] The first wave of industrialization began in Europe in the late eighteenth and early nineteenth centuries with the mechanization of the cotton industry, the use of coal for smelting metals, and the application of steam power. The leading sectors of steel, chemicals, and electricity were developed in the mid and late nineteenth century. This was the time of the so-called "Second Industrial Revolution." The use of more complex technologies has led scholars to characterize these industries as "scientific" because they were supposedly based on new scientific knowledge.

England led the world in the cotton, railroad, and iron industries until the latter half of the nineteenth century, when competition increased and England's industrial rivals began to catch up. In the new industries of chemicals, electricity, and steel, Germany and the United States quickly overtook England in output, and went on to dominate world markets. In his account of this process Mathias states that the electrical industry was probably the most important of the new industries, and the failure to develop this sector as quickly as foreign competitors "proved to be one of the most strategic weaknesses of the time and one of the most potentially damaging for the future."[2]

The literature on the electrical industry in England and its performance up to the time of the First World War has been almost unanimously unfavorable. The great economic historian Sir John Clapham has called the lag in electrical technology the most "deadly" lag in new technology.[3] Hughes noted a "considerable and persistent" lag in the British supply industry, well behind that of Germany and the United States.[4] A common theme of literature on electrical development is the unfavorable comparison of England's electrical industry with those in Germany and the United States. For example, in his survey of electrical development up to 1914 Leslie Hannah stresses the German excellence in the electrical supply industry and criticizes "the patchwork of differing standards" which prevented standardization and consolidation in the English industry.[5]

Electricity supply statistics have been used to prove the existence of a lag. The fact that there were many more supply systems in the United States than there were in the United Kingdom in the 1880s has been cited as evidence of a lag. Hutchinson argued that by 1888 every American town of over 20,000 inhabitants had an electricity supply while "central power stations were a rarity in Great Britain." This was a lag that came into existence "despite pioneer electrical triumphs of Faraday."[6] Most authors begin their description of the rise of the electrical industry with the experiments of Faraday, but the beginnings of the "electrical age" are usually dated from the 1880s, after the introduction of the incandescent lamp in 1879. Similarly, the lag in English technological development is dated from the 1880s, when it appeared that German and North American electricians had taken the lead from England's electrical pioneers. Textbooks on European history typically date both the electrical age and England's lag from the 1880s.

The dating of the electrical age from the 1880s distorts the actual situation. Compared with the developments of the 1890s, electrical technology in the 1880s was primitive. Many authors have mistaken the first steps of electrical development as the final article. Furthermore, the descriptions of the lag in electrical technology tend to be static rather than dynamic, and rarely distinguish between stages of electrical development. It is stretching a point to say that a lag existed in "central power stations" in England in the late 1880s, as Hutchinson has done, when the concept of a central power station was not fully developed until the 1890s. The speed of electrical development in the nineteenth century meant that lags were appearing and disappearing at a rapid rate, requiring an analysis that is dynamic rather than static.

Evidence of the lag has been discerned in three areas: the backward and disorganized supply industry; the weak English electrical manufacturing industry, dominated by foreign firms; and the slow progress of factory electrification. The backward state of the supply industry has already been mentioned. The weakness of the manufacturing industry has attracted great attention. According to Murphy "the extent to which an authentically British electrical industry existed at all was slight" in the period up to 1914.[7] There was no English company as large and as influential as the great German and American electrical manufacturers like Siemens or Westinghouse. The literature on electrical engineering argues that large companies are quicker to innovate. Theoretically, the larger company has more resources to make capital investments which incorporate innovation, and it can take advantage of economies of scale. The development of formal research and development laboratories within the larger electrical manufacturers has supported the notion that applied science was the key to success in the so-called "scientific industries". In his account of the English electrical industry Richard Tames argues that the work of the "lone gifted amateur" was replaced by the "systematic data produced by trained teams of research workers," and that "innovation and growth lagged most in those industries where science played a part."[8]

Foreign companies made an impressive showing in the leading markets for electrical goods. During the 1890s the number of foreign subsidiaries in England increased, and by 1900 the manufacture of electrical equipment in Great Britain was dominated by foreign firms and foreign capital. The large traction systems built by American and German companies are held up by Mathias as examples of the superiority of foreign technology.[9] The presence of large foreign firms has often been taken as evidence of backward technology and engineering in the home industry. A Japanese scholar, Takuji Sakamoto, has presented an interpretation of the lag in terms of the dominance of foreign companies. His argument is that English companies were not able to keep up with the R&D of their German and American competitors and thus were unable to match them in developing new technology. The English companies could not keep up with the rapid development of new technology and consequently were roundly beaten in the market place. In Professor Sakamoto's view, English leadership in electrical technology "came to an end" in the 1890s when more complex, and scientifically based, technology was introduced.[10]

The role played by foreign companies in English electrical development has often been misunderstood. There can be little doubt that they dominated the market for specific goods during specific time periods. There was, however, no continuous domination of the electrical market in England by foreign companies. The work of the R&D organizations was not solely responsible for the lead that German and American companies had in more advanced technologies. The concept of science-based electrical technology has clouded the issue of technological change rather than clarified it.

The accumulation of scientific knowledge played a minor role in the development of new electrical technology. Much of the work of the early research and development organizations was concentrated on problem solving in existing systems rather than in generating new basic knowledge. In a penetrating article on the relationship of science and technology, de Solla Price has gone so far as to minimize the importance of scientific knowledge in industrial technology. He argues that "the trouble is that it seems very difficult to put one's finger on any incident where new scientific knowledge leads to an invention and then to its practical innovation."[11]

The area where the larger companies excelled was in perfecting and promoting technological systems. Electrical technology was not developed in the form of major individual inventions. It can be found in a mass of minor inventions and improvements. The progress of electrical development was dependent in part on "learning by doing", on the accumulation of many different engineering skills, and on the skills of the work force. It was in the format of large electrical systems that the larger companies were able to take full advantage of their resources and personnel.

Empirical studies of modern industrial technology do not support the contention that larger firms are quick to innovate. In an examination of eighteen major industrial technologies Stephen Davis found that small companies were quickest to innovate. Once an innovation had been successfully introduced into an industry larger firms were then quick to adopt.[12] These findings correspond to the development of nineteenth century electrical technology. The large multinational corporations rarely produced the initial invention or innovation. They were able to dominate the industry by purchasing or acquiring inventions and then diffusing them rapidly.

It will be argued in the following chapters that the large foreign companies owed their technological leadership to the depth of engineering skills they could apply to a technological system, and not to their command of new scientific knowledge. Furthermore, their superiority in two critical areas of electrical development--production engineering and marketing--owed little to the industrial research laboratories. Although electrical technology is understood to be science-based, its application required engineering skills. New electrical machinery had to be adapted to suit specific working conditions and to fit into an existing pattern of use. Similar problems attended the application of steam power. As William Ashworth has pointed out, "innumerable improvements were called for in the machines that it drove.... Use of new and improved machines created fresh problems as fast as it solved old ones."[13] Electric power systems were more complex and larger than steam power systems; consequently these kind of problems occupied the attention of research and development organizations. Very little basic scientific knowledge was sought by these laboratories. On the other hand, the solution to pressing engineering problems often had important spill-overs into areas of scientific interest. One expert on R&D organizations concluded that until recent times R&D departments contented themselves with engineering problems and left scientific research to universities and research institutes.[14]

The marketing of electric power has largely been ignored by historians of electricity, yet it played an important part in the diffusion of electrical technology. Electricity was an entirely new product on the market and methods of pricing it and selling it had to be devised at the same time that methods of generating and distributing it were being perfected. Marketing electricity involved creating a price structure that would suit existing methods of generation and educating the public into accepting radically new forms of light, heat, and power. It will be shown in chapters 7 and 8 that the large foreign companies took the lead in marketing and used advanced marketing techniques to dominate the sale of electrical goods and services. The lag in marketing was an important one, and it will be argued that once the lag in marketing was reduced after 1900, English companies began to make up the ground lost in the previous decade.

The lag in industrial power has been interpreted as hindering the increase of productivity and competitiveness of English industry. This argument follows from the premise that electric power represented the "best practice" technique at any given time. Although this kind of reasoning is very good for those with an interest in selling electrical power (and was probably created with such an end in mind), it does not help historians in creating an unbiased picture of the use of electricity in English industry. Ensor has argued that factory electrification indicated best practice technique, and that "this grew slowest in Great Britain." He cites a "rabble" of small, inefficient supply undertakings producing expensive electric power, which cost between 6d and 8d per Board of Trade Unit (one kilowatt hour). On the other hand, it was believed that the foreign factory owner had the benefits of cheap electric power. Ensor maintained that only the large, well organized supply systems of Newcastle and Manchester could match the foreign price of around 1d per Board of Trade Unit.[15]

The belief that the English supply industry was an inefficient and disorganized body of mismatched elements is a common one, and it has been blamed for many of the ills that beset the industry, including the slow adoption of electric power. The use of different voltages and electrical systems has led more than one scholar to the conclusion that electricity supply in England was inefficient and expensive, and that this played a key role in the slow electrification of factories. Parker and Reid note in their outline of economic history that "little use was made of electricity in English industry" and that "output was insignificant compared with Germany and the United States." They argue that this was because the English system was inefficient due to the use of several voltages.[16] The statistical analysis of the German, American, and English supply systems in chapter 2 will examine their development.

Another reason often cited as having a deleterious effect on the English electrical industry is the amount of public control of the supply and generation of electricity. The unique legal framework created to govern electricity supply in Great Britain has often been blamed for producing a lag; the Electric Lighting Act of 1882 is supposed to have retarded growth in the industry by placing it in the hands of municipal government who were unfit to manage a high technology industry.[17] The Act of 1882 did have an important effect on the evolution of the English supply industry, but its major failings are not to be found in giving the municipalities control of electricity supply.

The context in which new technology is introduced plays a vital role in determining the character of that technology and the speed with which it is diffused. Sahal has pointed out that "the long term development of technology is governed by the characteristics of the larger system of its use."[18] This system or environment is comprised of existing industrial techniques and economic and institutional factors. The economic factors governing technical change have been studied by economists such as Zvi Griliches and Edwin Mansfield. Griliches has suggested that the rate of adoption is determined by the profitability of the technology and the cost of the process. The logistical curve has been found to best represent the pattern of adoption of new technology over time. At any given time, the number of non-adapters deciding to innovate will be proportionate to the number of adopters who are already using the process. As this latter group grows, due to increased information and competitive pressure, the "band wagon" effect occurs as it becomes less of a risk.[19]

Yet these economic variables do not act in a vacuum because institutional factors can affect the profitability of the technology and the cost of adopting it. For example, government regulation of overhead wiring in the English traction industry greatly increased the cost of adopting electric traction. In his account of the development of electric traction, McKay has stressed the part played by institutional factors in slowing the speed of diffusion. He does not interpret this slow diffusion as economically damaging. He makes the point that every new technology needs to be incorporated into a framework which allows the consuming public to have some input as to the character and extent of the new technology.[20]

Chapters 4 and 5 will follow this lead. The Act of 1882 will be assessed in terms of the influence it exerted on future development. The rise of regulation will be a major concern of these chapters because regulation had a direct effect in increasing the price of adoption and decreasing the profitability of electricity. A comparison of the institutional environments of Germany, England, and the United States will show how different institutional frameworks affected the rate of diffusion. Institutional factors slowed down the rate of diffusion, but on the other hand, the English public had the benefits of safer systems of electric light and traction. Different institutional frameworks produced different environments for electrical technology and this is a major factor in the dissimilarity of diffusion rates. This process will be shown to have had the most impact on the formation of large systems of electricity supply which were the forerunners of the national grids. Government policy and municipal politics played an important role in the slow growth of a national grid in England.

Criticisms of inept government policies have often gone hand in hand with criticisms of poor entrepreneurial performance in accounts of the English electrical industry. For example, Hennessey concludes that the pattern of English adoption of new technology was "a long tale of sluggish enterprise, missed opportunity and Gibertian muddle." He interprets the dominance of foreign companies as a process of foreign engineers bringing in technologies where "Britons apparently feared to" and the slow growth of electrical usage as "an indictment of Edwardian enterprise, or rather lack of it."[21]

The pattern of English adoption of electrical technology can be interpreted as an indication of poor entrepreneurship only if that technology offered significant advantages over the light, traction, and power sources already in use. The low cost and wide availability of competitive sources of light and power made electricity less attractive to English entrepreneurs than to their German and American competitors. Furthermore, the inappropriateness of some electrical technology to the English context is not fully recognised in the literature of the lag. A case in point is the diffusion of three-phase alternating current technology. This form of alternating current was best suited to large supply systems based on hydroelectric power. In Germany and the United States, three-phase systems were developed to bring light and power to industrial areas. In England there was little hydroelectric power and industrial users preferred to operate their own power plant. The inappropriateness of three-phase technology therefore played a role in its slow diffusion in England.

The failure of certain forms of advanced current technology to be diffused in England led some scholars to conclude that engineering and scientific expertise was insufficient to support them. This has been part of the view that the lack of scientifically trained personnel lay behind the slow development of the so-called "scientific industries" in England. Cardwell has interpreted the loss of industrial predominance as more a failure in science and technology than an economic weakness.[22] Levine concludes that a major cause of economic stagnation lay in the neglect of applying science to industry.[23] Landes has argued that there is a "direct and evident" link between technical education and industrial progress, and his account of European technological development stresses the superiority of German technical education.[24]

The poor state of English technical education has been a major theme of accounts of late Victorian technological development. Bennett notes that England's industrial competitors often had the advantage of better educated workers, especially in the chemical and electrical industries.[25] The failure to adopt advanced electrical technology has been interpreted as the failure of English engineers to keep abreast of scientific progress. Byatt has concluded that English electrical engineers did much better in a period of relatively uncomplicated technology (1880s) than they did in a period when advanced technology was being introduced (1890s). In the key area of three-phase technology, Byatt noted that English electrical engineers were apathetic to its possibilities.[26]

An important characteristic of electrical technology was its international scope. The basic scientific and engineering knowledge of electricity moved across national boundaries and across the Atlantic. The fraternity of electrical engineers was international: engineers moved from one country to another with ease. Within the international framework of electricity, the arguments for national technical education lose a lot of force. The multinational companies had some positive effects on English development. They provided skilled engineers, they disseminated new technology, they financed its introduction and they trained generations of new engineers. This study will emphasize the positive contributions made by foreign companies in England.

It cannot be argued that there was no lag in electrical development in England during the period 1880-1914. This study will concentrate on assessing the parameters of the lag and explain why the English pattern of diffusion was different from those in Germany and the United States. The causes of the lag will be sought in the context of economic, technical, and institutional factors. It will be argued that the contemporary perceptions of the lag were exaggerated and that the reality of electrical progress was much different from the picture of backwardness created in the late Victorian era, which has subsequently permeated into modern accounts.

CHAPTER TWO

THE DEVELOPMENT OF ELECTRICAL TECHNOLOGY AND THE ENGLISH ELECTRICAL INDUSTRY

The development of electrical technology can be broken down into at least four phases, each a distinctive period in the evolution of the modern system of generation and distribution of electric light, power, and heat. The appearance of the first generator in 1831 marks the beginning of the technological development of electricity, and this important event was preceded by a long period of scientific investigation into electricity and electromagnetism, that went back as far as the ancient Greeks. The electrostatic generators of the late seventeenth century provided the basis for experimentation with the new force. The invention of the Leyden Jar in 1745, and the voltaic pile in 1796, gave experimenters a reliable source of current and made possible an enormous increase in the knowledge of electricity. The voltaic pile (a type of electric battery) constantly renewed its charge and it had the further advantage that many piles could be connected to one another to produce very strong currents.[1] This feature enabled scientists to experiment with the illuminating and heating qualities of electricity.

The first phase of electrical technology followed Faraday's discovery of electromagnetic induction. This phase embraced the design and improvement of generating machines, the invention of the electric telegraph, and the commercial development of the first electric lighting apparatus. The first phase ended in 1879 with the invention of the incandescent bulb. The second phase of electrical development occurred in the 1880s when the first form of electrical supply was conceived and the component parts assembled. This phase marks the beginning of the supply of electricity to the public. The perfection of public supply and the introduction of electric traction and motive power are the characteristic features of the third phase, which lasted from 1890 to 1900. The fourth phase took place in the first fourteen years of the twentieth century. This phase saw the application of electric power to industry on a wide scale, the electrification of mainline and underground railways, and the merging of supply systems into large grids. Although political considerations prevented the creation of national supply grids, most of the relevant technology was available by the end of the First World War. The twentieth century also marked the emergence of electrical technology for mass consumption, in the form of consumer durables and the accompanying marketing skills.

All four phases of the development of electrical technology were international in character. It will be shown in Chapter 3 that many conceptions of the lag began with the premise that electricity was discovered by English scientists. The popular belief was that Michael Faraday invented the dynamo and electric motor and Sir Humphrey Davy discovered the arc light. The contribution made by Englishmen in the first phase was no greater than those made by Frenchmen, Italians, or Americans. England was not the home of electrical science, nor could any one nationality claim pre-eminence. The starting point of Faraday's search for a magnetic source of electrical energy was provided in 1820, when a Danish professor of physics showed that an electric current produced magnetic effects. Hans Oersted's discovery was one of the most important in electrical science, and a pamphlet describing his experiment was widely distributed.[2] Davy and Faraday obtained a copy of the pamphlet and replicated Oersted's experiment. In 1821 Faraday found that a conductor carrying a current would rotate about a magnetic pole. This phenomenon had been described by Oersted, and Faraday's experiment illustrated an effect of electromagnetic rotation. This famous experiment was the first conversion of electrical into mechanical energy. Much of the theory about this electromagnetic force had been advanced by the French scientist, Ampere. But as Faraday was a "mathematical illiterate," he was not able to understand Ampere's mathematical theories.[3]

Ten years of sporadic and unsuccessful experimentation followed the initial breakthroughs of 1821. The work of foreign scientists again provided the starting point of Faraday's famous experiments of 1831 in which he induced an electric current by moving a bar magnet through a magnetic field. This discovery of electromagnetic induction was the reversal of Oersted's effect: converting magnetic force into electric force.[4] In one experiment Faraday replaced the forward and backward motion of the bar magnet in a hollow tube with the rotative action of a copper disc moving between the poles of a magnet. A current was induced across the disc.

This was the precursor of the dynamo, yet it was far from being a practical generator of current. The copper disc produced eddy currents which made the machine highly inefficient. Faraday called this device a "new electrical machine" and suggested that the weak current might be increased, "and probably by several arrangements." He did not explain how these improvements might be carried out or indicate what form they would take, but he was assured that "the latter would find their full development hereafter."[5]

A considerable amount of development had to take place before the dynamo suggested by Faraday reached a practical form. Faraday had shown the ways of producing electricity from magnetism, but there still remained the problems of arranging the conductors in the most efficient configuration and increasing both the magnetic field and the current it produced. Several new arrangements of conductors and magnets were produced in the decade after 1831, notably the magnetoelectric machines built by

Saxton (U.S.), Pixii (France), and Clarke (England).[6] These machines were constructed for use in the new field of electro-medicine which was fashionable from the 1830s onwards. The development of more powerful machines was given impetus by the idea of generating electricity for arc lights. A French telegraph manufacturer built the first of this new generation of dynamos. The Alliance Company used the design of Jean-Antoine Nollet, and the first commercial installation of an arc light and generator set was made at the South Foreland Lighthouse in 1858.[7]

The size of generators increased throughout the 1850s and 1860s because more current was needed to illuminate arc lights. Magnets were made much larger, but this failed to solve the problem of increasing the strength of the magnetic field. The important scientific discovery of self-induction was made by Joseph Henry in 1832, and this was followed in the 1840s by suggestions that the heavy and cumbersome permanent magnets of existing dynamos be replaced by electromagnets. The fact that electromagnets were incorporated into machines simultaneously invented in England, Germany, Denmark, Hungary, and the United States is further testimony to the international character of electrical development.[8] The current generated in these improved dynamos was not suitable for commercial use because of current fluctuations caused by poor armature design. The first important innovation in armature design was the work of the Italian scientist Antonio Pacinotti. His slotted ring armature of 1860 was an important addition to electrical knowledge, but unfortunately it remained in the form of scientific knowledge for ten years. The information that could have advanced electrical technology languished in the pages of an obscure scientific journal, unnoticed by the engineers who were trying to make a better armature.[9] This episode illustrates the importance of the process by which basic knowledge is transformed into technology and then put to use.

A Belgian inventor and entrepreneur was responsible for the application of the idea of the slotted ring armature and its rapid international diffusion. Zenobe Theophile Gramme patented a dynamo in 1870 which had a winding similar to the one designed by Pacinotti. Gramme made sure his machine obtained the widest possible publicity and he brought out a full range of commercial dynamos which could be used as either generators or motors. Gramme machines were designed for many uses, including electroplating and arc lighting.[10] The era of commercial lighting had begun.

Gramme machines were manufactured in large numbers and dominated the market for generators during the late 1870s. The widespread publicity given to these machines and their commercial availability increased the momentum of dynamo development. Gramme machines were vehicles of diffusion of electrical technology because they were bought, modified, and copied by inventors all over the world. There are several pages in Thomas

Edison's laboratory notebooks concerning work on a Gramme machine which was quickly imported into the Menlo Park laboratory.[11] Gramme's design became the basis for the generation of machines brought out in the 1870s and 1880s.

The improvement of the dynamo continued on an international basis. The next generation of dynamos was the work of engineers and inventors who were employed in the telegraph industry. The accumulation of technical knowledge and production skills involved in telegraph technology played a vital role in the development of dynamos and arc lights, for the "learning by doing" was transferred from the telegraph industry to the higher voltages and larger machines of electric lighting. The major step forward in dynamo design was the drum winding in which all the conductors in the armature cut through the lines of magnetic force. This innovation was the work of Hugh Hefner-Alteneck, chief engineer of the firm of Siemens and Halske, which was the leading telegraphic company in Germany.[12] The drum armature became the basis of a highly successful and widely distributed dynamo.

The best modification of the Gramme machine was made by Charles Brush, an American telegraph instrument maker. The Brush machine was larger, more powerful, and robust. In tests carried out at the Franklin Institute by Professors Elihu Thomson and John Houston, it was found to be better suited to the requirements of arc lighting. A complete arc lighting system was patented by Brush in 1877.[13] The Brush system became the most popular form of electric light in both Europe and the United States, where it was widely adopted for street and shop lighting. One of the first Brush installations was made at the John Wanamaker store in Philadelphia.[14] The major commercial competitor of the Brush arc light was the system designed by Paul Jablochkoff, a Russian employed by the French telegraphic firm of Breguet. The Jablochkoff candle was made in large numbers by Breguet, who also manufactured the Gramme and Nollet dynamos. One of the first arc lighting installations in Europe was made in the Avenue de l'Opera in Paris in 1878, and this was followed by several installations in London.[15]

The year 1879 saw several more installations of arc lamps in English provincial cities. At this time there were no English manufacturers of electric lighting equipment and consequently England became a valuable export market for electrical equipment manufactured in France, Germany, and the United States. The important scientific discoveries made by English scientists were not matched by equal endeavor in the engineering field, because technological leadership was in the hands of foreign engineers and companies. The leading electrical manufacturer in England in 1879 was the English branch of the Siemens and Halske company and the Berlin works of the parent company was the source of many of the dynamos used in European installations.

The first phase of electrical development ended on 21

December 1879, when Thomas Edison demonstrated his incandescent light and opened a new era of electric lighting. The key electrical technology of the 1880s was that of central station lighting. The concept of lighting large areas from a central source was developed into a practical form, and the major figure in this important advance was Thomas Edison. Although he is remembered as an inventor of the incandescent lamp, Edison's major contribution to electrical progress was his conception of a complete system of incandescent lighting in which many customers were supplied from a central generating point. Edison believed that the commercial potential of the new light could be found in concentrated areas of demand served by a single central station. The central station system involved selling the service to customers in an area of a few square miles and then linking them to a central source of current. The realization of this plan involved the design and construction of special dynamos, lamps, cables, switching gear, and fixtures. It also encompassed the development of an entirely new element in electrical technology: the central station.

Edison and his assistants worked night and day from 1880 to 1882 on this enormous project. The Pearl Street Station became the symbol of the new electrical era and was watched closely by entrepreneurs and the gas companies. The "learning by doing" that was involved in this work became the basis for much of the electrical technology of the 1880s, because everything had to be built from scratch. Many mistakes were made and many lessons were learned. Most of the problems encountered were mechanical rather than electrical in nature; for example, there were great difficulties in running dynamos and prime movers together.

The world's first central station opened in 1882 and the electrical world watched and waited to see if this pioneering installation would be a commercial success. The Edison Company was unwilling to develop the central station concept until the Pearl Street experiment had proved to be both economically and technologically viable. As it took several years for the first central station to make a profit, generating stations did not appear in any numbers until the mid-1880s. The technology developed at Pearl Street was quickly diffused in the United States and Europe. In 1885 there were 31 Edison stations. In 1887 there were 185 of them, each supplying about twice as many lamps as the earliest stations.[16] The Paris Electrical Exhibition of 1881 was the great showcase for the new Edison lighting system; one of results of this famous exhibition was that agreements were made to set up Edison companies in France, England, and Germany.

Although the incandescent lamp offered many advantages over the arc lamp in terms of quality of light, the arc lamp market expanded rapidly in the 1880s and a profitable business was found in street lighting and in the illumination of public places. In the United States the buoyant demand for electric lighting encouraged the formation of many small manufacturing companies, and by 1884 there were between 35 and 40 separate incandescent and arc light systems on the market.[17] Although the Brush Company was the early leader in the arc lighting field it was soon challenged by the Thomson-Houston Company, which was founded by the two men who tested the Brush system in 1878. The expansion of the Thomson-Houston Company illustrates the rapid rate of growth in the American electrical industry. In the first year of operation (1883) the company installed five arc lighting systems. By 1885 this figure had increased to 75, and at the end of the 1880s there were over 500 installations.[18]

The striking increase in American lighting installations impressed many visitors who crossed the Atlantic to review electrical developments. The pace of the diffusion of electrical technology in the United States can be gauged by statistics produced by the National Electric Light Association. This organization estimated that between 1886 and 1888 arc lamps in use in the northeastern United States increased by 913 per cent.[19] These figures indicate that the market for incandescent lighting was growing at a faster rate. Most of the lighting plant was in the form of isolated installations in which the consumer produced his own current. The Edison Company dominated this market. In 1881 there were eight installations; by 1883 this number had increased to 221 and it reached 702 in 1886. There were over 1200 Edison isolated plants by 1888.[20]

The decade of the 1890s, and the third phase of electrical progress, began with a boom of central station building in Germany and the United States. This was the result of technological improvements in lighting equipment which decreased the price and increased the quality of the light. In all three countries under discussion there had been a wave of central station building in the second phase, albeit on a small scale and with a technology that was soon obsolete. The small arc light stations of the 1880s were supplanted by small direct current stations modelled on the Pearl Street Station, which in turn were replaced by much larger stations in the 1890s. The establishment of lighting companies in every city, small and large, became the springboard for a second wave of construction which gathered momentum as the 1890s progressed. This process was most evident in the United States, where the large lighting market in the northeast encouraged the establishment of many supply undertakings. For example, by the end of the 1880s there were more than twenty companies supplying current in Philadelphia.[21] The formation of supply undertakings was financed by the leading electrical manufacturers, who were growing at a rapid rate. By 1890 the three major companies, Edison, Westinghouse, and Thomson-Houston, were responsible for about 1350 central stations and the employment of over 11,000 men.[22]

German supply undertakings also increased dramatically in the 1890s; new central stations were built and existing systems of public lighting were enlarged. In both the United States and Germany, the central stations built in the 1890s had larger generating sets and supplied many more customers. In these two countries the growth of electric lighting occurred in the first years of the 1890s, because most of the major city governments had made the decision to electrify public lighting by 1890.[23]

The introduction of electric lighting systems in England began shortly after the Paris Electrical Exhibition of 1881. There was slow progress until the latter years of the decade when many central stations were built. After 1888 the quantitative advance of lighting plant increased at a rapid rate. The expansion of English supply systems came a few years later than the German, for it was the mid and late 1890s before all cities had reached a comparable state of development. The reason for this lag is to be found in the institutional environment. The Electric Lighting Acts of 1882 and 1888 placed public supply in the hands of the municipal governments and these bodies were unwilling to innovate until electrical technology had been proven commercially viable. The pioneer municipal undertaking in England was begun in Bradford in 1889. The progress of this installation was watched closely by other municipal governments who were considering the adoption of electric lighting. The Bradford undertaking proved to be technologically successful but did not make a profit until the mid 1890s, and this delayed the wave of imitation.[24]

In terms of the number of central stations and their production of electricity, England lagged behind Germany. By 1893 German kilowatts connected in supply systems was roughly double the kilowatts connected in England. The exploitation of the lighting market in large cities provided the basis for the rapid expansion of German and American electricity supply. In England, many large provincial cities did not adopt the electric light in large numbers until the mid 1890s, which retarded the overall growth of the supply industry. The municipal governments of large industrial centers, like Liverpool and Manchester, were slow to adopt electricity because they had an interest in an existing gas works and they believed that electricity was a poor substitute for gas.[25] After a slow start, the municipal supply systems in England grew at a faster rate. Most of the new municipal supply works were built in the provinces and it is the increase in provincial connections which provided the major part of the boom in central station building in the late 1890s. The development of public supply after 1896 was to be largely the result of the great expansion of municipally-owned provincial undertakings. The number of kilowatts connected in the English supply system more than doubled from 1893 to 1896, and tripled between 1896 and 1899. By 1899 the German and English supply industries were almost equal in kilowatts connected.[26]

The electrical technology of the third phase was the technology of heavy electrical engineering. The developments of the 1890s were centered on the construction of large central stations, the extension of supply schemes, and the introduction of electric traction. The Edison central station system represented the highest point of electrical development in the 1880s. Yet its basic design was obsolete compared with systems developed in the late 1880s and introduced in the 1890s. The problem with the low voltage direct current used in the Edison system was that it was extremely expensive to expand the distribution system to supply more customers. An Edison station, or one modelled on it, could only supply current to a small area of no more than three square miles. Beyond this distance the copper dielectric of the cable had to be increased to an unpracticable and uneconomical size. The high price of copper determined the economic limits of the direct current system.

The means to expand supply systems was found in the use of alternating current which could be transmitted for long distances along small diameter cable. Alternating current could be transformed from low to high voltage, transmitted at high voltage, and then transformed down to low voltage at the point of use. This property led to a revolution in the design of central stations and distribution networks, for it made possible the very large supply systems which characterized electrical development of the 1890s.

English engineers and entrepreneurs played a major role in developing alternating current systems and marketing them on an international basis. The first system placed on the market was the Gaulard and Gibb primary transformer apparatus, which was manufactured in England and used widely in Europe. Gaulard and Gibb equipment was exported to the United States, where it formed the starting point for American research into alternating current.[27] The diffusion of early alternating current technology went on at a rapid rate in England during the second phase. Equipment was produced by several English manufacturers and the designs of William Mordey and Gisbert Kapp were turned into popular lighting systems. By 1890 there were about twenty supply undertakings in England using alternating current, including installations in London, Newcastle, and Sheffield.[28]

The most striking development in alternating current technology was the discovery of multi-phase alternators and motors. This occurred in the late 1880s and was first marketed in the early 1890s. Multi-phase current could be transmitted over very long distances and was well suited to drive motors. These were the two weak points of single-phase current. The decade of the 1890s, and the third phase of electrical development, opened with an impressive demonstration of the advantages of multi-phase current. In 1891 an international exhibition was held in the German city of Frankfurt, and the power and light for the exhibition was obtained at a hydroelectric plant near the village of Lauffen, about 180 kilometers from Frankfurt. The three-phase current was generated at Lauffen and transmitted at very high voltage to the exhibition where it was transformed down and used for a variety of purposes. The Frankfurt-Lauffen transmission brought worldwide publicity for three-phase technology and began an international wave of diffusion.[29]

A similar demonstration of three-phase current was made at the Columbian Exposition held in Chicago in 1893. This was followed by the great Niagara Falls project which became the largest hydroelectric installation in the world. A distinguished panel of electrical engineers decided to employ three-phase current to distribute electrical energy over a wide area, including the industrial region centered around Buffalo and Toronto. The massive size of the equipment used at Niagara Falls and the large area it provided with electricity gave excellent publicity for the three-phase system.[30]

These pioneer installations were followed by many smaller undertakings in Europe and the Americas. Three-phase systems were ideally suited to exploit sources of hydroelectric power, bringing light and power to areas which had previously been without a cheap power source. Consequently this technology was popular in developing nations where coal supplies were scarce or expensive. Three-phase installations were made in Italy, South Africa, South America, and Australia. Extensive use of this technology was made on the Pacific coast of North America, where there were good sources of hydroelectric power and poor availability of coal. Electricity was generated in the mountains and distributed among mining centers, industrial areas, and urban communities in California, Colorado, Oregon, and British Columbia. The first successful hydroelectric installation after the Niagara project was made at the mining center of Telluride, Colorado.[31]

The great advantage of three-phase current was the reliability and high efficiency of the three-phase induction motor, which made the system ideal for industrial use. In many cases the source of electricity was the waste heat of coal mines and steel plants which could be harnessed to raise steam. A large integrated concern could generate electricity at several widely dispersed locations and then distribute it over the total area of the works. This kind of industrial power system was begun in Rhenish Westphalia and Silesia, where there were great concentrations of industrial activity. The largest systems were the Rheinisch-Westfaliche Electrizitats A.G. (RWE), and the Elektrizitatswerke A.G. which both supplied the industrial needs of the Ruhr.

The first public supply of three-phase current in England was not begun until 1899, well after similar systems had been started in Europe and the Americas. This pioneer undertaking was the work of the Newcastle Electricity Supply Company (Nesco), which changed over to multi-phase current in the late 1890s.[32] Nesco was unique among English supply companies in that it actively courted industrial users within a large geographical area. The lag in introducing three-phase current to supply networks and industrial operations was acutely felt in England in the 1890s, and it became a major component of the perceived lag in electrical development. After Nesco's successful introduction of three-phase supply, several other supply companies decided to create large multi-phase schemes. Unfortunately, the institutional framework of public supply worked against the formation of large schemes and there was slow development of multi-phase undertakings in England during the third and fourth phases.

The lag in electric traction in the third phase of electrical progress was the most obvious sign of an overall lag, because the electric tram was a highly visible manifestation of electrical progress. To the great majority of Europe's population the first indication of the great potential of electricity was the replacement of the horse tramway with electric traction. The introduction of the electric tram in the 1890s marked a serious lag in England, for it was foreign engineers and companies who dominated the market and diffused the new technology. The lag in electric traction was particularly disturbing to the English engineering community because many of the foundations for the new technology had been laid in England. Most of the early research into the possibilities of electric traction had been carried out in Great Britain. Yet most of these pioneer traction schemes were small and catered to the leisure market, such as Magnus Volk's light railway on the Brighton seafront. Several power sources were tried in urban transport schemes, but steam power, cables, and compressed air failed to replace the horse. As soon as electric motors were available, they were applied to the tram and there were several experimental tramways in Germany, Great Britain, and the United States. These all failed because the electric motors were not sufficiently robust and powerful, and there were continual difficulties in the pickup of current.[33]

The introduction of multi-polar generators and motors provided the solution to the problem of increased power output and power-to-weight ratios. One of the leading motor designers in the world was the American engineer, Frank Sprague, who formed his own electric traction company and agreed to electrify the horse tramway of Richmond, Virginia, in 1887. He admitted that his traction system was based on "largely underdeveloped and untried ideas," and he bore the brunt of heavy criticism from the engineering fraternity.[34] The crucial problem of current pickup was solved by the invention of the overhead trolley pickup, which was begun by the Belgian inventor, Charles Van Depoule, and then perfected by Sprague. After the electrification of the Richmond tramway opened in late 1888, contracts were made for 200 systems of electric traction in the United States, 133 of them with the Sprague Company; 90 per cent of them employed the Sprague system.[35]

The 1890s began with the rapid adoption of the Sprague system of electric traction. The leading American electrical manufacturers were quick to acquire the personnel and patents required to market the new technology. The Edison interests acquired the Sprague Company and the Van Depoule system was bought by the Thomson-Houston Company. These two multinational organizations then sold traction schemes on an international basis. Germany was the first European country to adopt the new traction technology, when the Bremen tramway was electrified by Thomson-Houston in 1890.[36] The two largest German electrical manufacturers quickly became involved in electric traction and Allgemeine Elektrizitats Gesellschaft (AEG) and Siemens made large profits in electrifying European tramways.[37]

The earliest municipal tramway schemes in England were in Leeds and Bradford, both of which opened in 1896. These pioneering undertakings were slowly followed by other municipalities in the last years of the 1890s and the first years of the twentieth century. In terms of electrified tramway construction, England lagged a long way behind Germany and the United States. In 1895 there were 45 route miles in Great Britain compared with 254 route miles in Germany and 1300 miles of electrified tramway in the United States. By 1900 the American figure had reached 19,314 miles, the German route miles now numbered 1800 while the English route miles had only reached 572.[38]

The traction boom in England came several years after similar waves of development in Germany and the United States. It went hand in hand with a boom in central station building that took place at the end of the century and attracted many new entrants into the English electrical industry. Up to this time English electrical manufacturers had remained few in number. The market for their goods and services was not large enough to support a large manufacturing industry during the second phase, and the increase in electrical activity in the early 1890s served to induce foreign manufacturers into the English market rather than to stimulate the home industry. Although several large electrical manufacturers immediately set up subsidiary companies in England to market the new systems of incandescent lighting in the 1880s, the lure of a large traction market in the early 1890s brought a major foreign penetration into the English market.

The growth of electric traction in the United States had an immediate effect on the technological and economic conditions of the industry. The large market for urban transport enticed the electrical manufacturers into the field and accelerated the process of product diversification that had begun in the late 1880s. Firms like Thomson-Houston rapidly expanded their product line and their operations as they became competent in new fields. The possibility of enormous profits in electric traction led the company to extend its operations to Europe in the mid-1890s, and set up manufacturing facilities in England.[39] The Edison group of companies continued to expand during the 1890s, acquiring several new companies, and setting up foreign subsidiaries. Both the operational and financial needs of creating traction schemes demanded a large enterprise with a centralized administration and this played a large part in the movement towards consolidation in the electrical industry of the United States.[40] The formation of General Electric out of the Edison companies and Thomson-Houston in 1892 is an example.

The establishment of foreign subsidiaries of the major multinational electrical corporations caused anxiety among English electrical manufacturers and consternation in the electrical and popular press. The "foreign invasion" appeared to reach a peak in 1900, when the market for electric traction, advanced three-phase equipment, and industrial power schemes were dominated by German and American companies. In stark contrast to the multinationals doing business in the United Kingdom, the English electrical manufacturers remained small and tended to specialize in a narrow range of products. In the early years of the 1890s English electrical manufacturers (Brush, Electrical Construction Company, Crompton) sustained a healthy growth but by the 1900s these manufacturers were dwarfed by the American multinationals (British Westinghouse, British Thomson-Houston), whose annual yearly output was much greater than English companies.

The foreign exploitation of the English electrical market is also evident in the importation of foreign electrical machinery into Great Britain. During the 1880s, much of the electrical apparatus used in English supply systems was obtained from abroad. Although English electrical manufacturers gained ground in the 1890s, a large trade deficit for electrical goods existed and reached a peak at the time of the "foreign invasion" of the twentieth century. Byatt has estimated that imported machinery accounted for 38 per cent of capacity installed in English power stations during the period 1896 to 1903.[41]

The beginning of the twentieth century, and the fourth phase of electrical progress, occurred in the middle of a boom in the English electrical industry which was sustained by the expansion of municipal lighting systems and the construction of joint traction and lighting schemes. The expansion of the municipal supply and the enlargement of the manufacturing industry went on until 1905, when the pace slackened. The boom in central station building and traction electrification went hand in hand with a great expansion of the electrical trade in industry. The fourth phase also marked the real beginnings of the electrification of English industry.

The new markets in industry and in consumer durables attracted many new entrants into the electrical manufacturing industry. Most of these new companies were English, and there were some major differences between them and the existing foreign subsidiaries. The newcomers did not offer a full range of products and services, but instead concentrated on the production of specialized equipment. The multinationals manufactured and marketed complete traction, public supply, and industrial power systems. This was their great strength in the electrical market, for they provided a package of goods and services that included the design, installation, and operation of electrical systems. In this way they dominated the market for the large installations that normally attracted public attention. The most visible and news-worthy electrical installations were traction schemes.

After a comparatively slow start in the 1890s, English electric traction undertakings increased rapidly in the first years of the twentieth century. The 572 route miles of 1900 were extended to 2,117 route miles by 1906, and in 1913 this figure had reached 2,595.[42] By 1906 the traction boom was over in England for most of urban systems were completed by that time. Two new areas of electric traction opened up in the fourth phase: underground railways in urban areas and electrification of mainline passenger railways. English engineers had been pioneers in both areas. The first electric underground railway in the world was the City and South London Railway. This was an English enterprise and it became the pilot scheme for many subsequent underground railways.[43] The first major mainline electrification in Europe was carried out by the North Eastern Railway. This project was begun in 1902 and involved about 80 miles of electrified line. The engineers for this scheme were provided by

Nesco, as was the current and all the electrical plant.[44]

Despite these early efforts, the two new areas of electric traction proved to be disappointments for the English electrical community. Although the City and South London Railway turned out to be a successful experiment, it did not encourage English investors into supporting and enlarging the underground railway system in London. The investors who finally put up the capital were German and American. At one point in the fourth phase, two rival schemes were competing for Parliamentary approval; one plan was financed by Charles T. Yerkes, the American tramway magnate; the other was underwritten by the J.P. Morgan group. Finally, the Yerkes plan was approved and the London underground was constructed with American equipment, designed by American engineers, and financed with German and American risk capital.[45] The London underground railway was a showpiece for electrical progress in the twentieth century and its domination by American interests brought a barrage of criticism in the press. Here was another area of electrical technology in which pioneering English development seemed to be surpassed by superior foreign engineering and financial talents.[46] The London underground railway was a highpoint in perceptions of the lag and a major element in the public's awareness of a "foreign invasion" around the turn of the century. The North Eastern Railway electrification was a great success, but the boom in mainline traction failed to materialize before 1914. It was not until the

1930s that any effort was made to electrify mainline routes.[47]

The English supply industry experienced steady growth during the fourth phase. New power stations were built to meet the increase in demand for electricity, and these "super" stations employed the latest boiler and turbine designs to produce very large amounts of electricity as cheaply as possible. Although there was a significant increase in production in the English industry, it still lagged behind the rapid growth of the German supply system. The expansion of public supply in England was carried out by the municipal undertakings and the privately owned bulk supply systems. These bulk supply organizations planned mass production of electricity, and they experienced steady growth despite municipal opposition. The promoters of the bulk supply companies expected to surpass the output of the municipally dominated supply industry during the fourth phase. Their expectations were not met because large municipal and private supply undertakings accounted for a larger proportion of kilowatts connected. The bulk supply companies were intended to supply electricity in bulk to existing supply schemes and provide greater standardization. This they were largely unable to achieve, for the supply industry was loath to surrender its independent generating capability. The debate over bulk supply in England provided much of the evidence used to depict a lag in electrical technology.

Standardization was a problem in the English supply network

because it had been constructed from many different pioneering systems of direct and alternating current. The many different voltages and frequencies hindered standardization and amalgamation, and gave the English supply industry the appearance of being a "patchwork" of small and inefficient systems. The electricity supply of London was the most striking manifestation of this situation. Over seventy separate supply organizations provided electricity for the metropolis, and each undertaking employed a different type of distribution and supply. The case of London's electricity supply was hotly disputed in the electrical and popular press during the fourth phase, and criticisms of the situation fueled perceptions of the lag.

The fourth period of electrical progress was the real "Age of Electricity," because in the early years of the new century the mass of the population had the opportunity to take advantage of electrical technology for the first time. The electrical industry became more consumer oriented, and its efforts were directed into the expansion of the market for electricity. The third phase had been one of heavy electrical engineering; of large supply systems and electric tramway schemes. This was the period of the greatest lag in England and also the time that the multinationals dominated the market. The emphasis returned to the consumer in the fourth phase, because the expansion of the large supply and traction systems was slowing down as the major installations were completed. In addition, the new traction technologies did not produce an increase in business that could be compared to the traction and central station booms of the 1890s. The electrical manufacturers turned their attention to the expanding market for light, heat, and power which was found in industry and in lower middle and working class districts. The emergence of technologies which greatly reduced the cost of wiring and connecting consumers' premises opened up large new markets in urban areas.

The most important new technologies of the fourth phase were in electrical goods and services for mass consumption. This was the foundation for the "Age of Electricity," when the majority of the population had cheap electricity to use in a growing number of consumer goods. There had been a continual process of "learning by doing" in the manufacturing industry which resulted in a drastic lowering of costs for lamps, switches, fixtures, and wires. A great deal of effort was put into improving the design and manufacture of incandescent lamps, because this was an area in which improvements could greatly decrease costs and improve the quality of light. The large multinational firms directed their research and development effort into finding better filaments for the incandescent bulb. In the fourth phase metallic filament lamps were introduced which used less current to produce a better light.

There was continual pressure on the supply industry to increase demand for its product because of growing competition from the gas industry. The development of superior incandescent lamps and the lowering of costs in all areas of public supply increased the number of people who were willing to adopt electric power and light. New markets were found and new methods were devised to sell more electricity. The introduction of advanced lamps brought more customers, but it did not produce a proportionate demand for current because the new equipment was much more efficient in its consumption of electricity. This increased the pressure on the supply industry to open up new markets and find more uses for the augmented amounts of electricity it was capable of generating.

Industry became one of the most important new markets for electrical goods and services in the fourth phase. The possibility of using electric power in industry had been realized in the early days of the electric light, but only the German and American electrical manufacturers had developed an industrial power market. During the third phase, only Nesco constructed the large industrial power systems which were common in Germany and the United States. The lack of electrical power in English factories became a major element in the perception of a lag in electrical technology. The factory inspectors, who were in the best position to estimate how much electric power was being used in English industry, often used their reports to comment on the low level of electric power in the 1890s. Around 1900 this theme changed to one of description of the striking increase in electric power usage.

In the 1901 report of the factory inspectors, they noted that "first and foremost in the development of industry is the extension of electric power," and concluded that "electricity will be as important in Great Britain in the twentieth century as steam was in the nineteenth."[48] The replacement of existing power sources by electric motors increased during the first years of the new century, with the greatest number of new users obtaining current from public supply mains. In 1903 the factory inspectors estimated that 200 of the largest supply undertakings were supplying 54,000 h.p. of electric motors and the total electrification of British industry was thought to be in the region of 100,000 h.p. of electric power. They estimated that 80 per cent of the increase in electric power came from public mains. The first users of electric power in industry had employed isolated plants to generate their own electric power. As bulk suppliers began to cultivate power customers and the large municipal undertakings started to promote the use of motors on public mains around 1900, there was a noticeable increase in power usage via large supply systems.[49] The factory inspectors estimated that from 1903 to 1904 the increase in industrial power usage was 57 per cent, and from 1904 to 1905 there was 40 per cent more electric power being used. By 1905 they estimated that British industry was using 300,000 h.p. of electric motors, and it was evident to them that isolated plants were on the increase and that they now formed an important part of the movement to install electric power in industry.[50]

Despite the accelerated development in electric power usage in English industry there was a lag compared with the amount of electric power used in Germany and the United States. Problems in marketing and applying electric power in England resulted in some industries using a great deal of electric power, and others using extremely small amounts of electricity. In engineering and shipbuilding, great progress was made in applying electric power with the result that by 1912 about 75 per cent of the power in use was electric. This figure compares very favorably with figures for the United States and Germany estimated by Byatt for the period just before the First World War. On the other hand, electric power in the textile industry was less than 10 per cent of total power used in 1912, which was considerably less than electric power usage in the United States' textile industry.[51]

The unsatisfactory progress of the electrification of British industry provided more evidence of the backward state of the supply system. The consequences of a backward supply system were ominous: expensive electric power would handicap English industry in the struggle for world markets. The findings of several government committees after the First World War reinforced this belief. The English electrical industry was viewed as a weak link the nation's industrial structure.

CHAPTER THREE

PERCEPTIONS OF A LAG IN ELECTRICITY

The discussion of a lag in electrical technology formed part of late Victorian England's concern with loss of industrial leadership. The last two decades of the nineteenth century witnessed great social, economic, and political changes in England; chief among these changes was the readjustment to a new phase of industrial technology. The "Railway Age" that had been marked by British commercial and strategic domination was being replaced by another age in which world markets proved to be much more difficult to dominate. The "Second Industrial Revolution" brought more competition to world markets. Britain was facing the economic challenge of its major industrial rivals, the United States and Germany. The lag in electrical technology was a visible example of backwardness in the new "Age of Electricity."

Although much of the literature on the lag treats it as a static phenomenon, the lag, and contemporary perceptions of it, were constantly changing. The lag in electrical technology grew and changed in time with the four stages of electrical development outlined in Chapter 2. Contemporary opinion, as manifested in the professional and popular press and in the writings and speeches of industrialists, statesmen, and engineers, often had a mistaken impression of the lag. Most contemporary accounts placed English leadership in the first phase and the lag in the second phase of electrical development. The third phase was the time when the lag seemed most striking to Victorians, and this perception of the lag reached a high point around the turn of the century. There can be no doubt that there were lags in the adoption of specific technologies in the 1890s, yet the elimination of these lags and the reasons for slow electrical development were not always appreciated.

The key to public interest in electrical technology is to be found in the popular belief that electricity stood at the pinnacle of scientific endeavour. Invisible, mysterious, and powerful: electricity personified the forces of a new age dawning in Victorian England. It also became the barometer of industrial performance in a new era of industrial technology. Cheap and plentiful coal was acknowledged as a major contributor to "England's commercial supremacy" and "the secret of our industrial success" by a contributor to the Victorian Magazine, Nineteenth Century. This writer correctly noted that electricity was going to be the equivalent of coal in the twentieth century, and that commercial supremacy would go to those nations who took full advantage of the new source of power.[1] This conception of the lag, and its consequences, is typical of the articles and speeches concerning electricity published in late Victorian

England. Electricity was the power source of the future, and the great concern with commercial performance led to a growing interest in new industrial technology.

This preoccupation with England's commercial supremacy came as a result of the changing world situation in the last twenty five years of the nineteenth century. It has been argued that the "Great Depression" never existed, and that only some sectors of the industrialized economies suffered, while real wages and living standards generally rose.[2] Yet a long period of falling prices and profits was accompanied by expressions of concern on the part of contemporary Englishmen. The Royal Commission appointed to investigate the "Depression of Trade and Industry" published some disturbing findings. It pointed out that the "increasing severity of competition of foreign countries" was eroding the economic position of the Kingdom and that "our supremacy is being assailed on all sides." The Report also acknowledged that the conditions under which England had maintained industrial leadership could not be expected to operate permanently and urged English industry to come to terms with the new situation.[3]

Increasing competition from Germany and the United States was initially evident in a small number of industries, principally engineering, iron and steel production, and chemicals. These were the industries that England had originated and mastered in the early stages of industrialization, but by the end of the nineteenth century these industries seemed to be the ones in which foreigners excelled. As The Times noted in 1899, "we have had a hard time keeping the business, when we once had a monopoly."[4] Public interest in English commercial supremacy had strategic as well as economic overtones. The Boer War gave notice that the imperial fabric was weakened and that the Empire had some dangerous new enemies. It also marked the end of imperialism. "Splendid Isolation" had lost a lot of its splendor by 1902 and British foreign policy had to be readjusted to a new strategic balance in the world.[5]

To Englishmen who recognized that military strength was positively correlated to industrial capacity, the advance of England's industrial rivals was alarming. Both Germany and the United States had benefitted from English entrepreneurship and technology in the early stages of industrialization. In the last two decades of the nineteenth century there were disquieting indications that the students had become the masters of industrial technology, especially in its most advanced forms of electricity, chemicals, and optics. Much of the literature of the lag in electricity appeared in comparisons with Germany and the United States.

Comparisons with Germany formed an important body of literature in late Victorian England as Englishmen sensed the threat of a "rival and successor."[6] A study of the industrial competition of Germany became one of the most widely read books on economics during this period. Made in Germany, by Ernest E. Williams, was dedicated to the manufacturers and merchants of the United Kingdom and was intended to shake them out of their "fatal torpor."[7] It outlined the reasons why German industry had made such impressive gains and pointed to the steps that had to be taken to avert England's decline. Among the advantages enjoyed by German industrialists were high tariffs, the helpful German consular system in foreign markets, and the excellence of German technical education. In order to survive in a more competitive world market, Williams advised English manufacturers to market their goods more energetically, to adjust to changes in trade and technology, and to keep their equipment up to date.[8]

These themes can also be found in the literature comparing England with the United States. The typical American industrialist was supposedly much more inclined to adopt new techniques than his English competitor. English writers had great respect for the "American system" of production which was supposedly based on advanced manufacturing techniques. American businessmen were often characterized as being more responsive to technological changes and more disposed to scrap obsolete equipment and replace it with advanced machines. Consequently their industries were supposed to be more efficient, more productive, and more profitable.[9] The loss of technological leadership in English industry acted as a unifying theme in the explanations of English industrial backwardness, for technological advances were believed to be a major factor in England's early lead in industrialization. To a "generation who had taken invention for granted," the relative decrease in technological superiority was intimately associated with the end of industrial superiority.[10] It was commonly assumed that inventive ability and a technologically skilled work force were the keys to industrial leadership and that "foreigners have the advantages over us in technology which has led to their commercial advantage." This typical viewpoint believed that the foreign worker was more productive because he was better educated and that technical education facilitated mastery of the newer, more complex, industrial technology.[11]

The slow development of the English electrical industry provided an excellent example of this theme of technological retardation. The lag in electricity was more often cited because it was more visible than the lag in chemistry. The lag in electrical technology was harder to stomach because it was erroneously believed that British inventors and businessmen had provided the foundations of the electrical industry and that this early lead had been lost in the 1880s. The heroic figures of Sir Humphrey Davy and Michael Faraday dominated the imperfect understanding of electrical development. A patriotic Englishman believed that the new light began with Davy and the dynamo was the result of the discoveries of Faraday. This ethnocentric picture of electrical progress is common to many of the contemporary accounts which appeared at the same time as the first electric light. A series of articles in The Economist in 1882, entitled "Modern Electric Lighting," began thus: "The electric light is not a new invention. Three quarters of a century have passed since the discovery was made by Sir Humphrey Davy..." The rest of the article places electrical development as the consequence of Faraday's discovery of induction.[12]

This viewpoint remained a characteristic of English attitudes towards electrical science during all four phases of development under consideration. In the mid 1890s the wife of Sebastian Ferranti made the following comment about her husband's business:

> It is an extraordinary reflection on our British laxity that later on we had to give up making dynamos entirely owing to the keen competition from abroad, after having practically taught the foreign makers how to build the machines.[13]

Mrs. Ferranti was incorrect in her assessment of the development of English dynamo manufacture. In fact, it was foreigners who taught the English how to manufacture dynamos. What the early English electrical manufacturers, like Ferranti, excelled in was the improvement of their copies of foreign machines.

In the first years of the 1880s electrical engineers and commentators in the professional journals had a highly optimistic outlook for the English electrical industry. Confident in their belief that English scientists and engineers had a strong grasp of electrical technology, these sources painted a very rosy picture of the future of electricity in England. They believed that electrical technology had reached a state of perfection (a serious error), and that electricity would soon be used to light many of England's cities. These optimistic feelings were encouraged by entrepreneurs who were trying to raise money for ambitious lighting schemes.[14]

This wave of optimism ended suddenly in 1882. That year saw the "bubble" of speculation in arc lighting companies burst in addition to the introduction of a key piece of legislation which was to exert great influence on the development of electricity in England. The Electric Lighting Act of 1882 provided the legal framework for the public supply of electricity, and was intended to protect the public from some of the dangers of electric lighting. The danger of speculation was all too obvious at the time because many investors were losing their money in the wave of bankruptcies in the supply industry. Another danger to be avoided was the monopoly of public lighting by private companies; the gas companies had given the public and their legislators conspicuous examples of the disadvantages of such a monopoly situation. Although the framers of the Act intended to strike a balance between public and private ownership of the supply industry, their efforts resulted in the balance being heavily weighted in favor of local government. A key clause of the Act was called the twenty-one year clause, whereby local governments could compel a private lighting company, operating in their jurisdiction, to sell the undertaking after twenty-one years of operation. The price of the undertaking was determined by what was termed its "scrap value," an exaggerated term which meant that the company did not get any compensation for the so-called goodwill it had built up during its years of operation.

The compulsory purchase clause in the Electric Lighting Act was inserted to enable quicker and cheaper municipal take-overs than had been previously experienced in the gas industry. The optimistic reports of the electrical industry, and the intense opposition to private exploitation of the new illuminant in the House of Commons, led to the original purchase clause of bill coming into effect after only five years.[15] The supporters of the electrical industry in Parliament had to make strenuous efforts to increase the period from five to twenty-one years. This process of negotiation was between the electrical industry, its supporters in Parliament, and the interests which considered the monopoly power of electric lighting as an anathema.

Perceptions of the lag in electrical technology began in 1882 and were a result of the debate about the compulsory purchase clause. Initially there was relief that the clause had not been made much shorter. Then it soon became obvious that there were to be no quick profits in electric lighting and subsequently the process of negotiation recommenced. The electric lighting industry strove to improve its legal position throughout the remaining years of the nineteenth century and for much of the twentieth. It had a lot to gain by vigorous lobbying in (and out) of Parliament. The regulation of the industry was opposed because it increased costs and reduced the prospects of profit. The promoters of the industry believed that the Act had placed electric lighting under too stringent regulation, and consequently their unbridled optimism was replaced by deep pessimism. The Act was supposed to have "strangled the baby at birth," and there followed a well choreographed campaign to bring public attention to this fact.[16]

The friends of the electrical industry were soon able to gather their forces to meet the threat of over-regulation and municipal takeover. The company promoters and their shareholders formed an influential block in Parliament, especially in the House of Lords. Opposing them, the gas industry allegedly had its interests well represented in Parliament and the municipal trading ethic had strong support in the Commons. It was believed that many prominent parliamentarians had interests in the electrical industry, including the leading Conservative politician, Lord Salisbury. Such were the rumors about Salisbury's connections with the electrical industry that he had to disavow that he was major shareholder in electric lighting stocks.[17]

The professional engineers within the electrical and telegraphic industry were in the process of forming themselves into national organizations when the Act was passed. They immediately began to organize their opposition to some of the terms of the Act. With the support of entrepreneurs and industrialists, the engineers became the vanguard of the agitation directed against the Electric Lighting Act. Eminent Victorian engineers such as William Preece, Silvanus Thompson, and Rankin Kennedy were men of stature and influence in English politics and society. They figured conspicuously as witnesses in Parliamentary commissions and committees. Their speeches were considered to be newsworthy and in them they helped keep the image of the lag alive in national consciousness. Soon after the Act was passed in 1882, a society was formed to oppose the Act and press for changes in some of its clauses, especially the twenty-one year clause. This lobbying group arranged meetings, public speeches, and a very large petition that was presented to Parliament in 1886.[18]

There was nothing to be gained in portraying the electrical industry as strong and healthy. The various interest groups who had joined together to amend the Act chose to depict the industry as weak and stagnating under ill-considered legislation. In 1882 the electrical industry felt the cold hands of government regulation around its neck and spent the rest of the century loudly threatening to expire. There was a great amount of publicity given to the retarding influence of the Act. The public's expectations had been raised high in the years before 1882 and there was some disappointment not to see electric lighting in the streets in the 1880s, after all that had been promised.[19] Although the real reasons for this situation were to be found in the primitive electrical technology of the 1880s, there was widespread belief that the English electrical industry had suffered from over-regulation. The Prime Minister, Lord Salisbury, spoke for many when he said that the Act of 1882 had "stifled" electrical development in England while its competitors moved ahead.[20]

It can be seen that the earliest depictions of the lag were based on comparisons with other, supposedly more advanced industries. The general process of comparison of England with Germany and the United States was immediately extended to the electrical industry. The perceived lag in English electrical development was based on a comparison with the advanced electrical industries in Germany and the United States. During the 1880s many electrical engineers and entrepreneurs visited the United States and the Continent to inspect electrical installations. This was a natural consequence of the process of imitation of electrical technology which was the basis for the English industry's development at this time. Engineers like William Preece travelled extensively in the United States, France, and Germany, and were given guided tours of the most advanced electrical installations. They returned to the United Kingdom with glowing reports of the advancement of electrical technology in England's industrial rivals. These impressions were published in both the professional and popular press and gained a wide audience.

American electrical development and engineering practices commanded a great deal of respect in English engineering circles. Much expert opinion in the English industry in the 1880s judged the potential of an electrical scheme on the yardstick of American practice.[21] In addition, there was a tendency for returning English engineers to exaggerate the pace of electrical development in the United States. Several accounts of American visits gave the impression that every community in the United States had an electric lighting system with its own central station.[22] This was not true. Some very large American cities had central stations, but most towns had a small arc light installation which was often represented in the English press as a central station. Many American communities installed arc lamps because they did not have gas lighting, a fact that was rarely mentioned by returning electricians.

Few could tell that the accounts of American development were often exaggerated. What mattered was the public's acceptance of the existence of a lag in English development. The public was certainly made aware of a lag from 1882 onwards. A few well-known figures were successful in using their position and their oratory to exert a powerful influence on public opinion. One such person was Lord Thurlow, who played a conspicuous role in the fight to reduce government interference in the electrical industry. Thurlow led the Parliamentary opposition to the Act of 1882, and continually publicized the damage the Act had done to English electrical progress. In speeches and letters he maintained that the institutional framework had made electrical development in England an "impossibility," and that it was only in England that this scientific endeavor was making no progress. Lord Thurlow promised that the repeal of the Act would promote electrical advancement and give the stimulus to a large industry which would absorb a great number of the unemployed.[23]

Lord Thurlow personifies the interest groups which had much to gain from the promotion of the idea of a lag in electrical technology. His depiction of a weakened industry and its great future potential suited the feelings of the times. His interest in these activities sprang from his financial involvement in the industry. He was a major shareholder in the newly formed supply companies, and was also a director of the Brush Electrical Company. In Parliament he was joined by politicians such as A.J. Balfour and Lord Crawford, who were also heavily involved in the new industry. In his speaking tours he was joined by men like Robert Hammond, a prominent figure in the supply industry and an outspoken commentator on the lag in electricity supply. Mr. Hammond was an entrepreneur who formed a number of supply companies.[24]

In 1888 the efforts of Lord Thurlow and his fellow agitators reached a successful conclusion. Thurlow's amendment to the

1882 Act was made into law by the new Conservative administration and the twenty-one year clause was replaced by the forty-two year clause.[25] Yet the agitation against legislation and government regulation did not end in 1888. In fact, it gathered momentum in the 1890s, for the third phase of electrical development brought with it new technologies and new threats of government control. The new technologies were in traction and long distance alternating current transmission, and both suffered from stringent regulation. The use of high voltage alternating currents and live tramway conductors proved to be the great manifestations of these fears. Both were killers. The most feared of electrical equipment was the overhead wire, which carried current in long distance transmissions and the electrical power for the overhead pickup of traction systems.

Described as the "newest danger of civilisation" in England, the overhead wire was considered to be the agent of electrocution and fire.[26] The alternative to the overhead wire was the underground conduit system. As this method was extremely expensive, electrical industries all over the world resisted government pressure to do away with the overhead wire. One way of combatting this pressure in England was to characterize all government interference as highly damaging to an already retarded industry. It was often argued by the members of the English electrical industry that they suffered from more safety regulation than any of their major foreign competitors. One of the reasons brought forward to explain the greater progress of the American and German industries was that they had less government interference to hinder them.[27] The amount of safety regulation was an issue which kept perceptions of the lag alive in England up to 1914.

In addition to the problems of government regulation, the tramway industry was saddled with a twenty-one year clause and the constant threat of municipal take-over. The poor comparative record of tramway electrification was normally blamed by the tramway companies on unfair and uninformed legislation. A leading figure in the fight against municipal and government interference in the tramway industry was Emil Garcke, who formed the largest tramway group in the country. Throughout the 1890s, his articles in the press and public speeches blamed slow development on government regulation. The twenty-one year clause was often cited as the reason why tramway mileage lagged behind that of the United States, Germany, and even Italy. Sir William Preece announced that because of the threat of municipal take-over, "the British financier has lost faith in electrical enterprises at home."[28] No reason was put forward to explain the eagerness of German and American financiers to invest in English tramway electrification.

The turn of the century marked the climax in contemporary perceptions of the lag. The great visibility of traction undertakings and the interest that the English public had in transportation helped to make the lag in traction a very prominent part of the overall lag in electrical technology. On the other hand, the lag in alternating current technology, which had also appeared in the 1890s, caused great concern to the engineering and scientific communities. The third period of electrical development involved the perfection of highly complex multi-phase machinery. New forms of electrical technology in the 1890s were based on a much more rigorous scientific foundation than earlier direct current and single-phase alternating current. The obvious lag in multi-phase current technology--there were no English equivalents of the Frankfurt-Lauffen and Niagara projects in the 1890s--caused many observers of the electrical scene to conclude that English engineering and scientific endeavor was backward. One article in the electrical press alleged that English engineers were apathetic to the potential of advanced alternating current technology and incapable of mastering it.[29]

Public attention was drawn to questions of electrical engineering competence because of the wave of publicity surrounding "The Battle of the Systems." This battle was the struggle between engineers and industrialists who supported the introduction of alternating current technology on one hand, and their colleagues who defended the existing direct current technology. The direct current interest were challenged by a competing technology and attempted to discredit it in much the same manner as the gas industry had tried to discredit the electric light when it appeared. The supporters of alternating current technology had to counter this publicity offensive with claims that their technology was the most advanced and that the direct current systems were obsolete. Comparisons with the more advanced alternating current systems were a useful way of indicating that England was behind in the electrical technology of the future. Thus in the press, in public announcements, and in speeches, the advocates of alternating current cited advances in Germany and America while criticizing the slow progress of advanced technology in England.[30]

The publicity given to this lag occurred at the time when interest and concern about English technical education reached a peak. The result was that supporters of advanced electrical technology and advocates of more technical education joined forces in the 1890s. Two well-organized and powerful movements to mould public opinion worked in unison to publicize the lag in advanced technologies and represent it as the consequence of poor technical education. The late Victorian age was called the "Age of Technical Education" by its most distinguished historian.[31] The movement to improve scientific and technical education was accelerated by the economic troubles and decline of English competitiveness in world markets. A speech by the distinguished scientist Silvanus Thompson, given in 1879, contains the prediction that "there is no question whatever that the persistent neglect of technical education in England will sooner or later ruin her in the markets of the world."[32]

The agitation for improvements in technical education occurred at the same time as the growth of the electrical industry in England. The Royal Commission on Technical Education lasted from 1881 to 1884, the Cross Commission sat from 1886 to 1888, and the Technical Instruction Act was passed in 1889. Further investigations and legislation continued into the 1890s. In 1897 the Royal Commissioners published another Report which stressed the superiority of German technical education and commented on the lead that Germany had built up in the scientific industries since the 1880s.[33] The perceived lag in advanced forms of alternating current formed a perfect example of this situation, and this example often appeared in the press.

Foreign penetration into the home market caused concern in England in the 1890s, and when a succession of very large electrical contracts went to American companies there was an immediate public reaction. In 1901 a speaker at the Institution of Electrical Engineers noted that "charges of backwardness are continually being made in regard to the applications of electricity in this country."[34] A best selling book in 1902 was called <u>The American Invaders: Their Plans, Tactics and Progress.</u>39 The fact that foreign companies were winning the largest electrical contracts was taken as proof of their technological superiority. Scant mention was made of their organizational, financial, and marketing advantages.

The discussion of the lag in advanced technology gave wide publicity to the supposed superiority of the German and American technology and in doing so increased their manufacturing companies' competitive edge in the market. The concept of technological superiority was constantly being reinforced by reports of municipal agents and engineers who were inspecting electrical installations in Germany, Austria-Hungary, Switzerland, Italy, and the United States in the 1890s and 1900s. These glowing testimonials brought back to England were of major importance in publicizing the lag, and increasing public perception of it. The municipal governments, who formed the largest element in the home market for electrical machinery, had always made a practice of visiting the leading electrical installations before making specifications for their own plant. Initially, their visits were confined to the British Isles, but as electrical technology progressed, these visits ranged farther afield. It takes little effort to imagine the reaction of a group of aldermen to the wonders of foreign technology while being feted in Germany or the United States. Many of them returned with an admiration of foreign plant and strong criticisms of the backwardness of the industry in England.[36] Engineering associations also took the opportunity to visit foreign installations and fraternize with their German and American colleagues. They were shown great hospitality and the newest, most advanced plant. The visit of the Institution of

Electrical Engineers to Germany in 1901 was a milestone in the formation of an image of German technological superiority. Reports made after this visit indicated that most of the work being carried out in Germany was in the most advanced forms of three-phase current.[37] This was an exaggeration. In fact, it will be shown that most of the new central stations being built at that time in Germany employed direct current systems. The electrical press often contained incorrect or exaggerated statements concerning the pace of foreign development or the low cost of electric power, which increased perceptions of the lag. When these statements were challenged, they were found to be erroneous; but they were rarely challenged.[38]

The superiority of foreign plant and the backwardness of the English industry became close to an obsession in the public mind at the turn of the century. The lag became a "self-fulfilling prophecy," because municipalities became so impressed with foreign plant that they framed their specifications in such a way as to give an advantage to foreign tenders. By restricting tenders to companies who had already constructed the specified machinery, the municipalities effectively excluded English manufacturers from taking part in the building of advanced technological systems.[39] Naturally the German and American companies, who were eager to increase their exports to England, did as much as they could to publicize their command of advanced technology. It was acknowledged by one electrical engineer that an area of German and American technological advancement was in their "intelligent and systematic advertising" of their technological superiority.[40]

The first years of the twentieth century did not bring about the expected triumph of the "American Invasion," for many fears of foreign competition in the electrical industry turned out to be groundless. The fourth phase of electrical development involved new technologies in which foreign superiority was much less evident. The press took more notice of large electrical installations like traction projects and super power stations than they did of imports of tungsten-element incandescent bulbs and consumer durables. After 1900, little was published about the lag in electric traction or in large power station machinery. The London Underground railway caused a storm in the press, but this project's obvious financial weaknesses soon reduced its value as a harbinger of foreign domination in the electrical industry. As technology changed the perception of the lag also changed. Many of the issues of the two previous lag periods remained, but they were organized around a different set of technologies.

The new form of public supply technology involved very large distribution areas in which large undertakings sought to sell electricity in bulk. These bulk supply systems were the first steps in a national grid of large supply companies. Here was a new chapter in electrical development and a new dimension in perceptions of the lag. The schemes for bulk supply met with considerable opposition; in fact, every existing supply organization opposed them when bulk supply bills were discussed in Parliament. At stake was the independent generating abilities of the existing undertakings. The municipal undertakings considered bulk supply a threat to their independence and to the privileged position that had been created by the Electric Lighting Acts of 1882 and 1888. What followed was a Parliamentary struggle which was carried on with great bitterness until the Great War forced a lull in action.

The bulk supply companies had to face the united opposition of the municipal undertakings, small private undertakings, gas companies and all their supporters in Parliament. The response to this entrenched opposition was a propaganda campaign designed to win the support of engineers, parliamentarians, industrialists, financiers, and the public at large. The bulk supply companies had a number of arguments that could be used to support their plans. They promised an efficient and rational system of public supply and a more economic means of generating and distributing electricity. They argued that their plans represented the rationalization and consolidation of the supply industry that was taking place in Germany and the United States in the fourth phase of electrical development. It was in their interest to make unfavorable comparisons with foreign supply networks and this they did in speeches, letters, articles, and

Parliamentary evidence. It was the bulk supply interests who brought the problems of non-standardization in the English supply industry into the public consciousness, and these perceptions of an inefficient and unstandardized industry still exist in modern accounts of electricity supply in England.

One of the most effective arguments used by the bulk supply companies concerned the provision of cheap power for industry. They supported the belief that English industry was unable to compete with foreign rivals because it was backward and inefficient. The bulk supply companies maintained that the source of industrial productivity was to be found in cheap power. They argued that the industrial rivals of England had the advantages of cheap electric power and that this gave them the competitive edge. It did not prove difficult to convince the nation that cheap electric power could be the panacea for industrial ills that was so fervently sought. It was asserted that the factory managers of England were desperate for electric power, and their viewpoint was reflected in sentiments like, "only bring it to our workshops at once, and at a reasonable price, and our effete old country will be rejuvenated."[41]

To support their claims, the promoters of bulk supply had to prove that electric power usage in England was far behind that of its industrial rivals. This proved to be a simple task. Their efforts in this area of the lag brought a great deal of attention to the backwardness in electric power and added this dimension to perceptions of the lag. The highpoint of the struggle to introduce bulk supply into London came when a number of companies attempted to reorganize the electricity supply of London. An enormous amount of data was gathered to show the backwardness of the existing arrangements and the meager use of electric power. Great efforts were made to convince the commercial and industrial community of London of the advantages of a rational supply system and cheap electric power. Every prospective power user was canvassed and massive petitions were organized. In spite of this large scale effort to influence public opinion the bulk supply bills failed to pass. Yet the agitation for the bulk supply of London had informed a large section of the population of the backwardness of the supply network and electric power usage, and convinced Parliament that steps must be taken to remedy the situation.[42]

Other forces were at work promoting the image of an industrial structure sagging because of an expensive and inefficient power source. Chief among these forces were the companies who had begun to specialize in factory and mining electrification. After 1900, several companies decided to concentrate on special equipment for electrification of mines, textile factories, and steel works. One of the marketing problems they faced was lack of interest of the factory owners and a general ignorance of the possibilities of electric power. Subsequently these companies were pushed into a program of public education, designed to increase awareness of the developments of electric power. Companies like Scott and Mountain publicized their own plant for mines and carried out surveys and comparisons of power usage in mines. Their results tended to stress the backwardness of English use of electric power compared with Germany and the United States and pointed to the great strides being made in these two countries.[43]

The activities of these varied interest groups produced a strong perception of the lag in electric power. There can be no doubt that a lag really existed, yet it was never as great or as long-lived as it was suggested by the propaganda of bulk power companies and specialized electrical manufacturers. There were several English industries with an excellent record of power usage, but their progress tended to get submerged in the general depiction of a backward and inefficient supply system serving industry that was unreceptive to new technology.

By concentrating public attention on specific elements of the lag and building public perception of these elements, electrical interest groups managed to cloud the reality of electrical development in England. The unsuitability of electric power for some sectors of industry received little or no publicity. The great progress in marketing electric light and power went unnoticed, and the considerable increase in municipal power circuits received no attention. While the perceived lag in traction dominated the 1890s, the completion of electric traction schemes in the next phase of electrical development did nothing to reduce perceptions of the lag because these perceptions were then centered on lags in power usage.

The perception of a lag in electrical technology served many different interest groups, and not all of them were connected with the electrical industry. The influx of foreign goods and the "American Invasion" was good propaganda for those pressing for the introduction of protectionism. The activities of municipal governments in the electrical supply industry provided ammunition for those groups dedicated to the curtailment of municipal trading. The lag served as a symbol of the economic and technological malaise that these diverse groups hoped to cure. The positive contributions of foreign companies operating in England were not stressed, nor was the municipalities' role in the extension of public supply. The real reasons for slow diffusion of some electrical technology in England were lost in the contemporary debates on technical education, municipal socialism, and the backwardness of English industry.

CHAPTER FOUR

THE INSTITUTIONAL BACKGROUND OF ELECTRICITY SUPPLY

The struggle between private and municipal interests to control the supply industry in England occasioned much of the publicity about the lag. The private companies, which had begun the sale of electricity in the first phase of electrical development, fought a losing battle against municipal ownership and government control. The response to this two-pronged assault was a propaganda campaign to bring public attention to the drawbacks of municipal control. The private supply companies argued that a great industry was being sacrificed "on the altar of socialism,"[1] and that municipal ownership was the cause of its backwardness, both in the amount of power generated and in the technology of public supply. Their efforts were successful in creating an image of a backward industry incompetently managed by municipal administrations. The Parliamentary Committee of the London County Council reported in 1907 that:

> It is common knowledge in electrical and financial circles, that technical enterprises such as tramways and electric light and power systems can be operated more economically and more efficiently under commercial than municipal management.[2]

This concept was "common knowledge" because the private lighting companies had spent twenty-five years publicizing it by

1907. The slow adoption of the technologies of central station supply in the second phase of electrical progress can be directly attributed to municipal conservatism. So can lags in advanced alternating current and traction technologies in phase three. Although some municipalities were slow to adopt new technology, there were several examples of enlightened municipal governments who vigorously expanded electricity supply. Bradford, Sunderland, and Sheffield, for example, were exceptions to the rule of conservatism and inefficiency in managing supply. Despite allegations of poor management, municipal supply systems could claim lower costs and prices than private undertakings. This issue was hotly debated in the electrical press. It was claimed by the private interests in the industry that the municipalities had lower costs because they could borrow money more cheaply. Their cheaper prices were said to be the result of their more lucrative market and the preferential treatment they received from the Board of Trade.[3]

Despite the publicity given to the backward state of the supply industry, successive British governments did little to prevent municipalities from taking over the lion's share of the lighting and traction market. The amount of municipal participation in the supply industry was substantially more than public ownership in the German and American industries. The amount of public control in England is responsible for many of the differences between English, German, and American development. Equally important in producing a different pattern of adoption was the great amount of regulation carried out by the British authorities.

The system of laws and institutions by which a society governs itself influences the diffusion of new technology. Electric light and power were especially dependent on a favorable institutional climate for rapid adoption. The practicalities of supply necessitates powers to break up the streets to lay cables, and rights of way over public and private property. The operation of central stations created nuisances, and the use of overhead wires led to fires and injuries. The novel and mysterious nature of electricity made it liable to regulation in the cause of public safety. It is doubtful that electric current from the first dynamos emitted dangerous radiation, as reported by the <u>Journal of Gas Lighting</u>.[4] It is equally unlikely that electricity could be used to cure all ailments known to mankind.[5] Yet to the nineteenth century mind all was felt to be possible, and legislators were as ignorant of the real nature of electricity as was the general public in the 1880s.[6]

The political climate in England in the 1880s was unreceptive to private control of the new technology. England had a government that was more experienced in regulating lighting than governments in Germany and the United States. Furthermore, municipal control of utilities, in both theory and practice, was most advanced in Victorian England. While Faraday was conducting his famous experiments into the nature of electricity, English politics was undergoing a struggle for a more representative form of government, which culminated in the Great Reform Act of 1832. Further acts of parliamentary reform were followed by similar reforms of local government as subordinate levels of government were brought into line with the new Parliaments.[7] This program of municipal reform was evident in most parts of the country, but nowhere was it carried out with more enthusiasm than in the city of Birmingham.

Modern Birmingham was the product of the industrial revolution. The city was dominated by the many factories and workshops which made it one of the most important centers of the European metalwork trades. The urban proletariat of Birmingham lived in appalling slums, and although there were continual calls for housing improvements, the municipal government had been reluctant to initiate policies that would increase taxation. Local government at this time was oligarchic and conservative, which made it totally unsuitable to tackle the problems of industrialization.

As Birmingham became more heavily industrialized, the plight of the working classes worsened, and movements for reform grew stronger. The leader of the reform movement in Birmingham was a successful industrialist who believed that municipal improvement could be achieved without excessive increases in taxation. Joseph Chamberlain's election as mayor in 1873 marked the beginning of a program of urban planning, sanitary reform, and slum clearance that was to be copied in England's other great industrial cities. Central to Chamberlain's administration was his takeover of the local gas company. He then ran the operation and plowed the profits back into the city treasury. Joseph Chamberlain considered the municipal gas supply in Birmingham to be his greatest and most influential achievement in public office.[8]

The supply of gas in England was a source of constant strife between the private gas companies and local government. The latter accused the former of selling inferior quality gas at inflated prices and making excessive profits. The English gas industry stayed prosperous and unrepentant despite attempts to regulate it. The continuing profits of the private companies increased municipal antagonism towards them, and relations between the two deteriorated. The municipal takeover of the Birmingham companies provided the means to end this strife. The new municipal gasworks soon proved to be extremely profitable, earning about $100,000 for the relief of the rates in the first six months of operation.[9] The great profitability of urban gas supply was to be a key factor in the program of municipal annexation of utility companies. The take-over of the Birmingham company was quickly emulated in other cities and it established a precedent which was followed by other local governments in the late nineteenth century.[10]

Local governments in Germany and the United States were faced with many of the same abuses of private ownership of gas supply. Yet as the development of municipal government was different in all three countries under examination, the institutional response to the question of public or private ownership of utilities was different.

The need to reform urban government in the United States did not become a major issue until after the 1860s, which was well after the movement had begun in England. In the late 1860s the poor governance of American cities had produced a "state of affairs no better, and much more difficult to remedy, than those laid bare by the Royal Commission of 1833 into British boroughs."[11] The 1870s saw a popular movement for urban reform which culminated in the downfall of several corrupt city governments, a curtailment of the spoils system, and new administrations in New York, Chicago, Pittsburgh, and St. Louis. The municipal reformers had promised extensive civic improvements, yet their programs rarely included the take-over of the gas supply. Several factors were at work in American local government which checked an American parallel of the wave of English municipal trading.

English municipalities enjoyed relative freedom from control by the central administration. The British government gave a measure of acquiescence, and in the field of electric lighting, positive encouragement, which was not evident in the United States. The movement for municipal trading in England coincided with high point of municipal self-government, but in the United States the question of municipal trading was brought up at a time when the cities were distrusted by the people and closely restrained by the state governments.[12] The corruption associated with public utilities was seen to be an important part of the spoils system and this made it difficult to gain state approval for municipal trading. As city government was considered to be an organ of the state government, the latter was able to interfere, especially in the financial aspects of municipal trading.[13]

The establishment of private companies in the gas supply industry came well before the initiative to reform urban government in the United States. Franchises had been granted to gas companies which gave them the right to supply for long periods of time and in some cases this right was granted in perpetuity.[14] As a result there was little public ownership of gas supply. By 1908 only about 25 of the estimated 800 American gas works were municipally owned.[15] The only municipal gas supply in a large city was at Philadelphia. This experiment in municipal trading proved to be an expensive failure and the operation was sold back to private interests in 1897.[16]

The municipalization of the gas supply in Germany was carried out at a greater pace than in the United States or England, for by 1908 65 of the 85 towns with a population of over 50,000 had their own gas works.[17] Of the fifty largest towns in both England and Germany, all fifty German cities had their own gas supply compared with twenty English cities. In England, 298 of the estimated 511 gas undertakings were publicly owned.[19] As soon as gas lighting was introduced into Germany there were attempts by city governments to gain control of it. There was a long tradition of German municipal trading and this soon asserted itself in relations with the new gas companies. The original franchises were of long duration and many of them incorporated purchase clauses which gave the municipalities the right to buy the undertaking.[20]

There was a great difference between German and American municipal trading, not only in the numbers of publicly owned undertakings, but also in the type of operation annexed by city governments. In 1911 a group of American councilmen was told in Berlin:

> You keep your sewers, which lose money, and give away your street railways which make money. You own your streets and parks, which are unprofitable, but not your gas and electric light works, which are profitable.[21]

American municipal ownership in gas and electric lighting was normally confined to small operations in rural or suburban areas. The average municipal plant was a small local operation whose costs, income, and profits could not compare favorably with the private sector.[22]

The German and English municipalities' participation in the gas supply industry was to provide important precedents when the question of electricity supply was raised. The activities of the gas industry in England strengthened the municipal reform movement to the extent that no private monopoly by a utility company was considered acceptable. To many local governments, the gas "monopoly" was the image of capitalism at its very worst and a prime target for municipal annexation. The words of Mayor Chamberlain expressed a belief widespread in local and national government by 1880:

> All monopolies which are sustained in any way by the state ought to be in the hands of the representatives of the people by whom they should be administered, and to whom the profits should go.[23]

The result of such beliefs was that comprehensive legislation to control a potential monopoly of utility companies was considered as soon as electric light seemed feasible. Thus in 1879 a parliamentary committee was formed to consider ways and means of providing proper control of the new illuminant. Although British government in the nineteenth century was supposedly dominated by the theories of "laissez-faire," there was a growing level of intervention in social and economic affairs. The amount of government intervention was great enough to convince some citizens that England was on the verge of socialism. Although what was essentially "gas and water socialism" was far from the ideals of Marx and Engels, municipal trading appeared to be growing in both size and scope. In 1880 municipalities believed that they had the right to supply electricity when it became a commercial possibility.[24]

The Electric Lighting Act of 1882 came as a direct response to the concern that private monopoly in lighting should be avoided at all costs. In the debates about electric lighting the major consideration of the government was that public bodies should be able to deal with electrical companies on better terms than they had with gas and water companies.[25]

The Act reflected the government's policy of preventing private companies from using their privileges to the detriment of the community, and it gave local authorities the first chance of establishing the supply of electricity. The framers of the Act also intended that it should enable limited experimental installations to be made and allow private supply when the municipality did not wish to exercise its option.[26] The many regulations concerning supply and the complicated process of applying for rights to supply were placed under the control of the Board of Trade. This organization was responsible for the implementation of the clauses of the Act, the granting of licenses and provisional orders to allow supply, and the supervision of those undertakings which were supplying electricity under powers granted in the Act.[27]

The Board of Trade had the powers to adjudicate between the claims of public and private interests, but in its early interpretation of the Act it always decided for the former. The license system, which was supposed to allow experiments, was not used for that purpose; in fact, very few licenses were ever granted. The main method of beginning supply was for the Board of Trade to approve a provisional order putting the right of supply under constant regulation and providing for eventual municipal annexation under the twenty-one year clause. Any application for a provisional order had to have the blessing of the local authority (in whose area supply was contemplated) before the Board of Trade would approve it. This gave the authorities the power to veto private supply in their jurisdiction. This veto power was used regularly in the 1880s and 1890s.[28]

The lack of municipal electrical undertakings, and their opposition to the extension of private supply, was often brought forward to explain the perceived lag in electrical technology in the 1880s and early 1890s. Private interests in the industry were outraged at the obstructionist behavior of the municipalities and accused them of deliberately retarding electrical progress. The depression in the electrical trades after 1882 was normally interpreted as the consequence of the compulsory purchase clause, which spokesmen from the industry claimed alienated capitalists and frightened off financiers. This popular theme has been taken up by many of the secondary sources and is best expressed in the words of a disgruntled managing director in 1890:

> Parliament has hampered and obstructed electrical lighting like no other. The reason why we haven't got electric lighting in England is simply because no one but an idiot would invest his money in an enterprise hampered by such conditions.[29]

On this surface this seems a plausible argument. The private investor had only twenty-one years to obtain a return on his money, and spokesmen for the electrical industry maintained that this was not long enough to recoup the capital outlay. Yet there had been few protests about the compulsory purchase clause when the Act was introduced. In fact, many sections of the electrical community thought that this clause would prevent further excesses in the company speculation which had severely weakened the industry in 1882.

There were slight prospects for a profitable supply industry in England in the 1880s. Electric lighting was often much more expensive than gas. The revolution that had been promised in electricity was not forthcoming because electrical technology was still in a primitive, and unprofitable, stage in the 1880s. This was at the heart of municipal obstructionism for they realized that electricity could not compete with gas as a public illuminant. The municipalities had little to gain by hindering the development of an industry that they were destined to control. Their primary need was for a cheap and reliable illuminant, and their expert advice often advised them that public lighting by electricity was not as good as gas.[30]

The slow entry of the municipalities into the supply industry was responsible for the unique pattern of electrical development in England during the second phase of electrical progress. While private companies built up the electrical supply in London, the municipalities in the provinces waited until the mid 1890s before embarking on the widespread introduction of electricity. The local governments chose the best electrical engineers in the country and heeded their advice to delay introduction of electric lighting until it was commercially feasible. For example, the municipal authorities in Bristol took an early interest in electric lighting and hired William Preece as their consulting engineer. Preece, an eminent engineer and electrical expert, advised his clients not to adopt electric lighting, as the technology was not fully developed and cost too much. This advice was given in 1884, 1885, and 1887. Municipal supply in Bristol was not commenced until 1893.[31]

One of the reasons why electricity was expensive in England was that safety regulations proscribed the often dangerous practice which cut the costs of electricity supply. The primitive nature of electrical technology in the 1880s made it an unsafe and unreliable form of lighting. Little was known about the fire hazards of electrical cables and the insulation required to avoid accidents. The majority of installations in the 1870s and 1880s were made with bare copper cables strung from telegraph poles and gas fittings.[32] The workmen who installed the first electric lighting systems were untrained and ill-equipped for the job. They were usually handymen, carpenters, or laborers.[33] Poor workmanship combined with ignorance of the effects of electricity inevitably led to fires and injuries.

Different measures to protect public safety were taken in England, Germany, and the United States. These measures were more stringent and comprehensive in England and they were successfully applied to the whole industry. One of the major aims of the framers of the Act of 1882 was to preserve public safety, for under its terms the Board of Trade was given the right to prepare and enforce "regulations securing the safety of the public."[34] One of the first regulations to protect the public from electrical dangers was the prohibition of overhead wires. When the idea of public supply of electricity was brought up, the municipalities made it clear to the Board of Trade that they were opposed to overhead wires. Any application for a provisional order that included the use of overhead wires was disallowed by the Board of Trade, and this excluded overhead wires for lighting circuits in nearly every urban center.[35]

Forcing the contractor to run wires underground was an important factor in raising the price of electricity because the technology of underground cables was in its infancy and their cost was from four to eleven times more than overhead cables. Many electrical engineers in the 1880s claimed that a practical and economic system of underground conduits had yet to be invented, and until it was, underground cables were costly, inefficient, and extremely difficult to maintain and repair.[36]

The danger of fire and the hazards of the overhead wire were also matters of concern in Germany and the United States, but there was a different institutional response to them. In the early 1880s, the number of fires and electrocutions brought bad publicity to the American electrical industry. Many accidents resulted from the widespread use of arc lighting in public places, which was a characteristic of American development in the second phase of electrical progress. The arc lamp used the highest voltage available in the 1880s, and as early lamps had no covering of the carbon points, sparks and pieces of red hot carbon fell from them. A great many fires were traced back to faulty arc lamps, including a serious fire in a Philadelphia textile mill which killed nine employees.[37]

The absence of a national body with the power to regulate

the growth of electrical undertakings compounded the dangers of electricity in the United States. This lack of central control was the result of the competing and overlapping spheres of authority in American cities. The ward bosses, the state legislators, the city councils, and the powerful mayors all had the power to approve or disapprove electrical installations. The lack of clearly defined powers was a hindrance to urban reform as well as a contributing factor to the spread of unsafe technology in American cities.[38] As there was no system of controlling applications to supply electricity, the American supply industry became a confused mosaic of private companies allowed to compete with one another with no regard to safety. In New York City there were 32 different companies running their wires around the city and some poles had over a hundred wires attached to them.[39] The results were described as "hideous." The wires were strung in a most casual and haphazard manner, which prompted one visitor to comment in 1899 that "everything has a temporary and expedient look about it which is very offensive to English eyes."[40]

There was considerable effort on the part of American local government to reduce the numbers of overhead wires. In this campaign they were aided by the press who warned the public of the potential dangers of them. Some municipalities fought the issue in the courts, but it can only be said that they hampered the use of overhead wires rather than eliminated them.[41] Some cities had more success than others in establishing some kind of order in electrical installations and removing the offensive wires. Strenuous efforts were made in Chicago, Baltimore, and Washington to control the use of wires and regulate electrical undertakings, and they met with some success. Yet it was a long and expensive process which did not attract a following in American cities and towns.

On the other hand, English wiring regulations were numerous and strictly enforced. In addition to the national rules made by the Board of Trade, each local authority had the right to make up specific rules for lighting undertakings under its jurisdiction. Local governments constantly monitored electrical installations and reported any infringement to the Board of Trade. The newly formed Institution of Electrical Engineers published a set of wiring rules in 1882, because electricians needed a comprehensive guide to wiring, switching, and connecting supply. The I.E.E. wiring rules stressed the need for safety and concluded that "the chief element of safety is the employment of skilled and experienced electricians."[42] It was customary for a consulting engineer to specify that these rules should be followed in the installation of electrical equipment. Yet these rules did not satisfy the insurance underwriters who were concerned with the fire risk of electricity. A wave of fires in the 1880s galvanized the insurance companies into action. Led by Mr. Musgrave Heaphy of the Phoenix Insurance Company, English insurers issued their own set of rules that had to be followed if the installation was to be insured against the risk of fire.[43]

All these rules were based on strict standards and were

constantly revised as electrical technology developed. Members of the electrical industry and the fraternity of electrical engineers often complained about the numbers of rules they were forced to observe. There was a consensus that excessive regulation greatly increased the cost of supplying electricity in England. One member of the Institution of Electrical Engineers commented, "if there had been an Institution of Gas Engineers and a set of rules, I do not think that there would be a single prosperous gas company in this country."[44]

In his book on the development of electric tramways in Europe, McKay notes that although Britain was a laggard in adopting this technology it was a "pioneer in her own right with a new institutional model for urban transportation" which was admired and copied in many European and American systems.[45] In the field of electric lighting, the slow adoption of technology for public supply in the 1880s was accompanied by the creation of an institutional framework for its regulation. Excessive regulation prevented the rapid electrical development that occurred in the United States, but provided a base for the boom in the English supply industry that came in the following decade. The slow growth of municipal supply in England in the 1880s was justified by the primitive technology in use at that time. Although exacting standards of workmanship and safety regulation increased the cost of electricity in England and slowed its diffusion, it also decreased the dangers of fire and injury. English electrical engineers often acknowledged that American development of electric lighting was going faster than England, but it was achieved at the cost of many more fires and electrocutions. One conclusion that could be drawn is that English electrical development in the 1880s was slower but safer.

CHAPTER FIVE

INSTITUTIONAL BARRIERS TO ENGLISH DEVELOPMENT IN THE THIRD AND FOURTH PHASES

The slow adoption of electric traction in England in comparison with Germany and the United States was a direct result of the distinctive institutional framework in which electric tramways had to develop. In all three countries the institutional framework was constructed at the time of the horse tram and then carried over in the third and fourth phases of electrical development. In Germany and the United States, the horse tram companies had obtained favorable franchises from city governments. The granting of franchises in the United States was normally accompanied by bribes and privileges given to the franchising body; consequently many of the franchises for tramways and electricity supply were of long duration; fifty or ninety-nine year franchises were common, and there were many franchises granted in perpetuity.[1] Frequently franchises did not specify a time limit, and court decisions normally interpreted them as perpetual or running for 999 years.[2]

In Germany, street car companies also obtained long-term franchises, but they were limited by precisely defined periods.

On average, German franchises did not extend beyond fifty years. The policy of municipal governments was to grant shorter term franchises as traction technology progressed. Therefore cities that had granted franchises during the horse tram era were often forced to buy out undertakings before the franchise had expired, and then electrify the tramway. Cologne, Dusseldorf, and Frankfurt all took over old established companies at great cost.[3] On the other hand, cities that granted franchises to electric tramway companies limited the time period and included provision for municipal annexation at the end of the franchised period. The franchise given to the Berlin tramway company is a good example of a German franchise. Under the terms of this agreement the company had to pay a fixed sum for the franchise plus a yearly payment. In addition, the company agreed to clean and maintain the streets where the tramlines were laid, and make additional payments if earnings exceeded a certain per cent of income. In the Berlin franchise the municipality was given the right to take over the undertaking, but the company successfully extended its operations beyond this time by appealing to the central government. Under the terms of the Prussian Light Railway Act, the government had the right to overrule municipal franchises. The Berlin tramway decision extended the company's franchise, but at great cost to the company.[4] This appears to be one of the few times that the central government intervened in a municipal franchise.

In England the operation of all urban traction schemes was governed by the Tramways Act of 1870. Under the terms of this Act municipal governments were allowed to construct tramway lines but were not allowed to operate them. The local governments could lease their lines to private companies for a period of twenty-one years.[5] This meant that when electrification was being considered in the early 1890s English tramway concessions had only a few years to run while continental franchises did not run out until the twentieth century, as most franchises were granted in the 1870s. While German municipalities bought out existing franchises or granted new ones for electrified traction, English local governments did nothing. The private companies operating traction schemes in England expected to be taken over at the end of the twenty-one year period and therefore did not expend great amounts of capital on electrification. For their part, the municipalities had the same wait-and-see attitude towards electric traction that they had for electric lighting. They were not going to risk tax payers' money on experimental projects which had a low expectation of return. They waited until their advisors told them that traction technology was both economically and technologically feasible before adopting it.

Another reason for municipal delay in electrifying tramways was the continual opposition to the overhead trolley wire system. The trolley wires were unacceptable because they posed the same threat to public safety as the overhead wire. English municipalities resisted the overhead system until all other alternatives had been tried and found wanting. Even when the conduit system had been proven more expensive and less efficient, leading local authorities like the London County Council (LCC) still insisted on conduits in the city center.[6] McKay has argued convincingly that safety and aesthetic objections were a potent force in delaying tramway electrification in Europe. He also points out that the American system of franchises and competing local authorities made it easy for the American tramway companies to avoid regulation and employ objectionable technology.[7]

The boom in electric tramways was late in coming to England. It was a municipal tramway boom because electric traction was in the hands of local government. Of the fifty largest towns in England by 1911 forty-two of then owned their tramway, while only twenty-three of the fifty largest German towns owned theirs.[8] Municipal ownership of American tramways was virtually non-existent in the nineteenth century. Only Detroit, Toledo, San Francisco, and Chicago managed to buy out the private tramway operations and did so at great cost. The private companies successfully resisted municipal annexation by fighting it in the courts and state legislatures. The result was that municipal ownership of tramways did not begin until the first decade of the twentieth century.[9]

The German tramway companies had less success in preventing municipal exploitation of urban transport. The franchises granted by local governments proved to be so detrimental to the profitability of tramway undertakings that the German electrical manufacturers, who sponsored the tramway companies, complained that the municipalities were bleeding them white.[10] Encouraged by the German government, the local authorities were able to use the profits from urban traction to reduce local taxes. On average, German municipalities made greater profits from tramways than English local authorities. A survey of the profitability of municipal trading in 1911 showed that the combined profits from gas, electricity, and tramway operations was three times more per capita in Germany than in England.[11] This indicates that municipalization of public utilities does not necessarily ensure higher municipal profits than a system of franchised private enterprise.

The German system of combining public and private ownership was also applied to electrical undertakings. Franchises were granted to private companies, but they were heavily weighted in favor of local government. High prices were paid for franchises, yearly payments were made, and most agreements between companies and municipalities included a compulsory purchase clause at the end of a fixed period. Clauses for maintenance and repair of streets were often included, and a schedule of payments was fixed to the amount of dividends or profits made by the private operations.[12] A spirit of cooperation existed between private and public undertakings that was absent in England. As electrical technology entered its third phase of development this difference was to have a significant effect of the progress of the English and German supply industries. The German system of coexisting public and private ownership was to prove more amenable to the adoption of technologies of the third and fourth periods of electrical progress.

The use of alternating current produced a revolution in public supply, as old direct current systems were replaced by, or integrated into, larger alternating current distribution networks. The major factor in retarding the development of large supply networks in England was the municipal opposition to them. The beginnings of this opposition came in the 1880s, when many local governments thought it was the most dangerous form of electricity. The "Battle of the Systems" brought the dangers of alternating current into the public eye because the direct current interests made this the central element in their opposition to the new technology. In the United States the supporters of direct current were led by the Edison interests, and the wide publicity they gave to the dangers of alternating current spread to Europe.[13]

The opponents of alternating current alleged that it was untested, uneconomical, and unsafe. The Institution of Electrical Engineers split into two parties, those who supported and those who condemned the new technology, but they all agreed that public discussion of the safety of alternating current could seriously damage the diffusion of all kinds of electricity.[14] The main advantage of alternating current was that it could be generated and distributed at high voltages, but this was the major objection to it. Local authorities, the Board of Trade, and the direct current interests all believed that higher voltages increased the chance of electric shock, electrical fires, and electrocutions. The use of alternating current to electrocute criminals in the United States reinforced these fears. As the institutional environment in England was highly sensitive to the dangers of electricity, regulations were soon made to limit the perceived dangers of alternating current. When the Electric Lighting Act of 1882 was passed the highest voltages in use were around 200 volts and this was made the upper limit of voltage allowed for public supply under the Act.[15] The Board of Trade was given the right to forbid the use of higher voltages and to regulate the insulation and protection of high voltage wires.[16] This power to regulate high voltage transmissions was to be often employed in the years after 1882, as voltages in common use increased. By the third phase of electrical development, voltages were being counted in thousands rather than hundreds of volts and during the fourth phase voltages of tens of thousands volts were being employed. Unfortunately, the regulation of high voltage transmission was too severe and it effectively retarded the development of higher voltage alternating currents in England.

In response to increased voltages, the Board of Trade made a distinction between high and low voltages and made separate regulations for each. The stringent regulation of high voltage cables was continually opposed by electrical engineers in the I.E.E., who claimed that the standards were too rigorous and that this contributed to the high price of electricity by limiting the economies made possible by high voltage transmission.[17] By setting the definition of "high voltage" lower than the technological norm, the Board of Trade effectively prevented the diffusion of high voltage technology. For example, if the feasible voltage of a project was higher than the Board of Trade upper limit, so many difficulties attended official approval of the higher voltages that the future and profitability of the scheme was placed into doubt. This problem was not confined to the supply industry only. Board of Trade regulations enforced a high voltage limit of 350 volts for tramway schemes when it was generally agreed, in the electrical industry, that 500 volts was a much more economical and satisfactory voltage.[18]

Following the lead of the Board of Trade, many municipal authorities did all in their power to hinder or prevent the employment of high voltage systems. The London County Council was conspicuous in its opposition to high voltage alternating current. As London constituted the largest single market for traction and light in the country, the opposition of this local authority had great impact on the slow diffusion of high voltage technologies. Believing that high voltages were a threat to public safety, the L.C.C. blocked several attempts to extend the supply of alternating current in London. It objected to the use of alternating current equipment, especially the street transformer which had a tendency to blow up, and it closely regulated companies who used alternating current.[19] When Ferranti tried to supply all of London with high voltage current in 1889 he met with resolute resistance from the L.C.C. Ferranti's Deptford scheme was totally undermined by the authority's refusal to have anything to do with high voltage current. Eventually a limited supply was arranged which employed high voltage cables, but the cables were only allowed on the private property of the railway companies that had lines running into the metropolis.[20] The L.C.C.'s opposition to alternating current was copied in several other municipalities because many provincial authorities looked to the L.C.C. for the lead in dealing with electricity. In this way, its obstruction of Ferranti's Deptford scheme provided the precedent for municipal opposition to the extension of supply systems and the distribution of high voltage alternating currents.

The Electric Lighting Act of 1882, and the municipalization of public supply, created an industry that was comprised of many small units. The direct current system was well suited to English institutional arrangements because it was economical within small distribution areas, which corresponded to the local government divisions of the country. Each municipal undertaking was independent and self-sufficient, with its own engineer, administration, and plant. The pioneering installation at Deptford undermined this practice because it was based on one main generating point and a large distribution network. When the question of bulk supply was brought before Parliament around 1900, the major point of contention between supply companies and municipalities was the future of the small municipal undertakings. While the bulk supply interests characterized those undertakings as unstandardized and inefficient, the municipalities appealed to the spirit of legislation in England which had proscribed monopoly powers in the supply industry.

At the time that the progress of electrical technology had made the municipal operation obsolete, the spirit of municipal trading was at its high point in England. In 1900 one of the first workmen to be elected to Parliament claimed that Great Britain was experiencing a renaissance of municipal administration which was the result of the growth in municipal enterprise.[21] John Burns was a socialist and an avid supporter of municipal rights, and his presence in Parliament indicates that new forces were beginning to be felt in national politics. The municipal electrical undertakings organized themselves in the 1890s. The Municipal Electrical Association was formed to disseminate information and technical assistance to municipal electrical works and to promote interchanges of knowledge within the publicly owned supply industry. It formed the basis for the organized opposition to bulk supply bills and also became a forum for municipal trading interests. It was the force behind the unified resistance to private monopolies in the supply industry. As soon as the intentions of the bulk supply companies had been made clear by the terms of the bills they promoted in Parliament, the Municipal Electrical Association immediately warned its members of a "flood" of bulk supply bills which threatened to create "one gigantic concern." It then began to prepare for a major confrontation in Parliament.[22]

While municipal trading and self-confidence had been rising, relations with the privately owned sector of the supply industry had suffered a serious deterioration. The activities of some of the more intransigent local authorities, such as the L.C.C., had offended and upset representatives of private companies in their jurisdiction. One spokesman for the private companies in London accused the L.C.C. of a "disgraceful and childish" opposition to the enlargement of London's supply network, which had deprived many Londoners of a supply of electricity.[23] The clash of interests between private and public undertakings came to a climax over the bulk supply bills.

The framework of institutional acceptance of bulk supply was laid by the Cross Committee on Electricity Supply of 1898, which recommended that powers be given to large undertakings to allow them to distribute electricity over large areas and sell it in bulk to existing undertakings.[24] The electric power bills that were to initiate supply in bulk were based on these recommendations. Yet these bills faced determined opposition in Parliament. The private interests in the supply industry calculated that the municipalities had a block of about eighty members of Parliament who were committed to oppose bulk supply bills.[25] Most of the electric power bills of 1900-1902 were passed by Parliament, but with important modifications. The municipalities had successfully campaigned to exclude themselves from the distributing areas of the bulk supply companies. This meant, for example, that the bulk supply company serving Lancashire was unable to sell electricity within the urban areas of Liverpool and Manchester, and the exclusion of the large municipalities from the supply areas was a great blow to the viability of the bulk supply schemes.[26]

The secret of success for a bulk power company was the acquisition and exploitation of large markets as soon as generating capacity had been installed. The bulk power systems had to be as large as possible in order to exploit economies of scale, and the large first investment and high running costs made early conquests of large markets vital for the successful operation of the scheme. Speedy salesmanship was crucial to the financial viability of bulk supply, as the first bulk supply companies were to find out.[27] The exclusion of the large urban markets, which were being supplied by the municipalities, made development of bulk supply very difficult and slow in England. The bulk supply companies were denied the largest markets and were therefore forced into developing new markets in industry and in rural and suburban areas. This serious disadvantage, combined with poor management and a developing technology of bulk supply, inevitably led to the financial discomfort and failure of some of the first bulk supply schemes. The South Wales Power Company, for example, was strenuously opposed in Parliament, denied all major urban markets, and eventually had to be reconstructed in a smaller form and refinanced. Much of its proposed generating capacity was sold to the municipal undertakings who had opposed it so successfully.[28] This example was repeated several times with the result that the bulk power companies grew more slowly than expected.

The large municipal organizations in England were slow to grasp the potential of bulk supply, but at least they came to the realization that large scale supply was the technology of the future and the small direct current installation was obsolete and inefficient. They admitted their mistakes in preventing bulk supply, but did not made amends to the private initiative they had restrained.[29] Instead, the large municipalities began to construct their own bulk supply networks based on their urban markets and the suburban and rural markets of their municipal neighbors. In this way Birmingham, Manchester, and Liverpool became bulk suppliers in their own right.

The only power company which was an immediate and thorough success was the Newcastle Electric Supply Company, which began bulk distribution in 1899. Nesco was aided by the local authorities within its area which allowed the company unlimited distribution rights. Municipal trading was not the policy of the Newcastle Council and thus the company was given a comparative freedom of action. The energetic leadership of Nesco and the high standard of its electrical engineering were the other major contributors to its success. The demand for Nesco's electricity grew at a rapid rate. In 1900 about one million units were sold and by 1905 this figure had increased twenty-five times. More power stations had to be built to meet this demand and the company successfully increased its distribution area. Nesco was the largest and most profitable bulk supply company in England. Its profits increased during the twentieth century and its dividends of around eight per cent were greater than average dividends paid by other supply companies.[30]

Nesco proved that good management and municipal cooperation were the indispensable ingredients for success in the modern supply industry. The experience of the German supply industry reenforces this conclusion. The supply of electricity in German was built on cooperation between private and public interests, whereas the English system was characterized by confrontation. The franchise system in Germany had forced a measure of cooperation between municipal owners and company operators of municipal supply networks. Many municipal authorities had become shareholders in the companies that had the franchise to supply electricity within their jurisdiction. This became the basis of the "mixed undertakings" which were to exploit the new technologies of bulk supply.

The organization of bulk supply schemes involved the amalgamation of many small distribution networks into one consolidated area. In Germany the impetus to form bulk supply networks came from the large electrical manufacturers who created the companies for that purpose. But rather than exclude the small municipal undertakings, they included them as partners in their schemes. The large distribution networks were jointly owned by the private companies and municipalities. The local authorities became the major shareholders in these schemes which gave them a measure of control and they also had their representatives on the boards of directors that ran the undertakings. In some cases municipalities gained a controlling interest in the private bulk supply companies.[31]

The formation and development of the Rheinisch-Westfaliche Electrizitats system provides an excellent example of a German "mixed undertaking." This vast company began with the municipal electrical works in Essen which was operated, under franchise, by a private company. A consortium of electrical manufacturers and steel producers, led by Hugo Stinnes and August Thyssen, took over this company and formed it into a bulk supply operation with the goal of supplying the power needs of industry in the Ruhr. This system was developed with full municipal cooperation and approval, with the result that the R.W.E. served about eighty towns and communes as well as the industrial giants of the Ruhr. Over forty per cent of the shares of this company were in the hands of the local authorities and they played an active part in its management. The close cooperation of municipal undertakings, which formed the lower level of distribution, and the large power stations, which generated most of the electricity, led to a highly advantageous grid system in which municipal stations provided "back up" capabilities to the main generation stations. The smaller municipal plants were called upon to meet excess demand when loads were high, and their large distribution networks provided an "electricity bank" of surplus power that could be called upon when needed.[32]

The state governments of Germany played an active and constructive part in the formation of "mixed undertakings." The state coal mines of Prussia and the hydroelectric plant of Bavaria brought these two governments into the business of bulk supply. The Prussian state government developed a system for the Saar industrial area utilizing very large power stations built on the coal fields. Municipal cooperation and joint ownership were the key elements in this system. In Bavaria the state government and the technical staff of the large electrical manufacturers planned and executed a large bulk power system which exploited both hydroelectric power and state owned coal fields. One of the main architects of this plan was the German engineer, Oskar von Miller, who was to become the leading advocate of a national grid in Germany. When the municipalities in Bavaria heard of this plan they objected to it fearing that a large scheme would deprive them of their independent generating capacity. The Bavarian government overruled their objections; the final compromise plan gave the municipalities powers of joint ownership, operation, and construction of the system.[33]

The bulk supply systems of Germany were the fruits of cooperation between private and public supply organizations; electrical manufacturers, industrialists, and large banks; and local and central government. Unlike the English bulk supply companies, they had a great deal of political clout at all levels of government. The association of German electrical engineers had influence in central government and used its lobbying powers with discretion and success. When the Prussian Post Office decided that its national control of telegraph and telephone system should be extended to electrical supply, it met with stiff opposition in the Bundestag. Leading German engineers and industrialists were active in a campaign to avoid any central control over the supply of electricity.[34] The ease with which the Prussian Post Office's initiative was repelled forms a telling comparison with the situation in England.

Events in England were to show that it was local government

who had the political power, and if the opposition of one local authority could block a bulk supply scheme by refusing to allow high voltage wires cross its territory, a united block of municipal interests in Parliament could overrule any political initiative begun by private supply companies. The success of the municipal campaign against bulk supply prompted private concerns to criticize a political system which gave so much power to municipal enterprise. The Chamber of Commerce Journal for April, 1900 contains a bitter account of the failure of a bulk supply scheme in Yorkshire. The municipal opposition to this scheme was organized on a national basis, and to the writer this represented "a combination compared with which the trusts of America were innocent in their oppression."[35]

The experience of the American bulk supply schemes illustrates how the different institutional environment contributed to a distinctive pattern of development. The English private companies had little influence in local and national politics compared with their municipal opponents. In the United States this situation was reversed because it was the private corporations who possessed the political influence, and they used it to promote bulk supply. English electrical engineers who toured the United States were impressed at the ease with which the supply companies could obtain political favors. One visitor noted that in England one had to get an Act of Parliament before commencing supply. In America, on the other hand, one only had to find the man who "represented" the local authority, "and the matter was settled easily over a long drink at the bar."[36]

The procurement of political favors and the exploitation of political corruption was part and parcel of running an electrical undertaking in the United States. Before Edison began his pioneering work at Pearl Street he had to court the local politicians and treat them to a sumptuous meal at his Menlo Park laboratory.[37] The buying and selling of political influence formed an important part of the relationships between supply companies and local government. It was a common practice to place prominent local politicians on the boards of lighting companies, and this gave the companies a measure of political influence and protection. Some lighting companies went to great lengths to secure the services of important city politicians. Political power was considered to be one of the most valuable assets of a supply company and it was alleged that a few companies in Philadelphia existed only to buy and sell it.[38]

The scale of electrical technology increased in its third and fourth phases and so did the amount of corruption in the supply and traction industries. The larger the undertaking, the more political influence was needed. Charles Beard saw this insidious process as leading to wide scale corruption in American cities. Writing in 1912, he argued that the supply and traction companies were keeping American cities "wide open." By this he meant that graft extended to all levels of the government and judiciary, and policies such as late-night tram service were adding to the general moray decay of the cities.[39]

The third and fourth phase of electrical development were the "Age of Consolidation" in the American electrical industry. A policy of consolidation and rationalization was first carried out by the electrical manufacturers and then repeated by the supply undertakings.[40] The technologies of bulk supply, which were developed in the third phase, were ideally suited to the consolidation of the supply industry. They proved to be the ideal vehicle to eliminate the smaller and less efficient units and to enlarge the powers and scope of the large supply systems. With no institutional barriers in their way, the large systems of New York, Chicago, Boston, and Philadelphia expanded into the suburbs and rural areas and annexed the smaller organizations that were supplying electricity there.

Samuel Insull's development of the Chicago undertaking is a good example of this process. Once Insull had acquired all competing companies within the Chicago metropolitan area, he began building power stations outside the city and used them to supply demand from urban and suburban areas. The existing suburban companies were incorporated into this large system as distributing stations for a supply area of about 200 square miles. Insull's next step was to buy the smaller rural supply companies and incorporate them into a state-wide distribution network which was organized in 1911.[41] The Boston Edison Company followed the same program. Charles Edgar's formation of a bulk supply scheme eliminated the fifteen smaller companies that were operating in the supply area. The distribution network was extended in suburban and rural areas and several very large power stations were built. By the fourth phase of electrical progress this system was meeting a level of demand that was not thought possible when the scheme was planned in the 1890s.[42]

The increased size of American supply systems brought them into contact with state governments, for distribution networks stretched beyond municipal jurisdiction. Very few state governments made any attempt to control or regulate bulk supply. In 1904 the state of Massachusetts considered a plan to establish a commission with powers over bulk supply. This commission had the right to review and revise all franchises and grants made to supply organizations. It had powers to fix rates, regulate cables, and supervise issues of stock. In a dispute between a private company and a local authority, the commission was to decide between them.[43] This idea met with little success for only Massachusetts, Wisconsin, and New York set up Public Utility Commissions beforè 1914.

In contrast to Germany and the United States, the English development of bulk supply in the fourth phase received some severe institutional checks. The struggle to create large, economical supply areas reached a climax in 1905, when a private company was formed to rationalize the electricity supply of

London. This system was chaotic and highly inefficient, symbolizing all that was wrong in the English supply industry. It also represented the bastion of municipal trading, for among the 72 different supply undertakings in London were some influential local authorities, including the L.C.C.

The plan to reorganize London's supply came from the Nesco company and it was financed by a group of investors and industrialists from the Northeast. The plan promoted by Charles Merz and his associates in 1905 was centered around a very large power station supplying electricity in bulk over the entire metropolitan area. Such were the economies of scale of the proposed system that it was estimated, by its supporters, to be able to reduce the price of current by seven-eighths.[44] The London County Council was informed by its experts that Merz's scheme was sound and that it was the best way to bring cheap power and light to London.[45] Despite this advice, the L.C.C. were determined that the bulk supply of London should not be in private hands and they led a stout opposition to the bill when it was introduced into Parliament. The opponents of the bill prepared themselves for a long and bitter struggle. Between thirty and forty learned council had been briefed to oppose the bill and the remaining five active members of the Parliamentary Bar were taken on by Merz. In all there were 77 separate petitions against the bill, with John Burns leading the opposition to it in the House.[46]

After a Parliamentary struggle of eight months the bill was ready to be passed but its opponents were able to delay it until the government fell and Parliament was dissolved. A fitting postscript to this unhappy episode was provided by Lloyd George, then at the Board of Trade. He was unimpressed by the technical merits of the plan and told Merz, "my dear young man, this is not a question of engineering but a question of politics."[47] Although Merz's plan was not put into practice, it had indicated to Parliament and the L.C.C. that further delays in reforming London's supply would damage the commercial and industrial life of the city. After Merz's funeral in 1940, it was said of his plan, "if the bill had passed, the (national) grid would have been put forward by almost a generation."[48]

There was to be little progress towards a national system of electricity supply in the remaining years before the First World War. The victory of the Liberals in 1906 brought many friends of municipal trading to power. Amongst them was John Burns, who took over the Local Government Board.[49] It was ironic that at the time American observers thought there was going to be a revolution in English bulk supply, it was in fact about to suffer its greatest defeat.[50] There was to be slow progress of the bulk supply companies, with the exception of Nesco, which became the largest supply system in Europe. Such were the delays and slow progress of the years before 1914 that a bulk supply scheme for all of London was not put into effect. It was not until the

1920's that Merz's plan was put into operation by the L.C.C. When the great power station was opened in Barking in 1925, it was built to Merz's design and stood on the site chosen by him in 1905. It provides a fitting monument to the delay caused by institutional opposition to modern supply technology in England.

CHAPTER SIX

THE LAG IN ELECTRICTY AND THE ISSUE OF ENGINEERING BACKWARDNESS

One of the implications of a lag in advanced technology is that the engineers and scientists of the lagging country are in some way deficient. It was commonly believed in Victorian England that the consequences of poor technical education were felt in the so-called "scientific industries." Charges of engineering and scientific backwardness were made in England during the lag and this theme is still present in the recent literature. The late Victorians believed that the early English lead in electrical technology was lost by the second phase and technological leadership then passed to the United States and Germany.

These contemporary perceptions were incorrect because England was a technical laggard--rather than a leader--during the first phase. The great progress made by English engineers during the second and third phases went unnoticed while the debate about technical education raged in England. The image of the poorly educated engineer who was unable to understand advanced technology provided good ammunition in the struggle to improve technical education, but was far from the truth.

The first electric lighting systems used in England were derived from French, German, and American technology. The designs of Gramme, Serrin, Jablochkoff, Brush, Edison, and Siemens dominated the market for electrical equipment. Although Joseph Swann invented an incandescent lamp, the manufacture and marketing of the lamp was carried out with the assistance of the Edison interests, and a joint company was formed in 1883.[1] The first English electrical engineers made equipment imported in the 1880s the basis for further development. One of the most important English electrical manufacturers started his business by selling imported Gramme dynamos and Serrin arc lamps. R.B. Crompton then redesigned the dynamo of the Swiss inventor, Emile Burgin, which in turn was an improvement of the Gramme machine. The Crompton-Burgin dynamo incorporated a compound winding of the armature which was an important technological advance. This innovation was the work of Crompton and his German assistant, Gisbert Kapp. It was patented in 1882.[2]

The international basis of the technological development of electricity is evident in the first two phases. Crompton saw the Burgin machine on a trip to investigate electrical installations in Europe. His assistant was one of the many young German electricians who were learning the trade in foreign companies. Gisbert Kapp attended a Technische Hochschule before joining the Crompton company.[3] Several German emigrants were trained in electrical engineering in Edison's laboratories and workshops in

New Jersey. Many of them, such as Johann Schuckert, returned to Germany to found large electrical companies.[4] English engineers and entrepreneurs made the most of this international network of ideas. They began the first phase of electrical development as laggards; most electrical equipment in England at this time came from abroad. By copying and improving imported machines, they made significant progress in diffusing the new technology in England. In the second phase of electrical progress, English engineers went from copying machines to the point of making substantial contributions to the theory and practice of electrical lighting.

The Edison direct current lighting system was the standard worldwide in the second phase of electrical development. Although it was an advanced system in the beginning of the 1880s, it was obsolete in 1890 when the third phase began. The dynamos used in the pioneering Pearl Street Station, and later imported to Europe, were heavy, inefficient, and prone to breakdown.[5] The 1880s marked some important improvements in the mechanical and electrical design of the dynamo. Better materials were used and more attention was given to reducing friction and sparking of moving parts. Yet until more was known about electromagnetism and the electrical workings of generators, there was little possibility of improving the low output and poor efficiency of existing dynamos. The great electrical engineer Silvanus Thompson wrote in 1886, "until we know the true law of the electromagnet, there can be no true or complete theory of the dynamo."[6]

The theory of electromagnetism and its mathematical foundations were provided by British scientists and engineers. The first great breakthrough occurred in the field of basic scientific knowledge. In 1865 James Clerk Maxwell published a paper entitled "A Dynamic Theory of the Electromagnetic Field," in which he outlined a mathematical model supplying the foundations of classical electromagnetic field theory.[7] This theory was derived from mathematical deduction and had no experimental foundation. As Hugh Aitkin has noted, it took 23 years to produce the experimental evidence to validate the model.[8] Yet much of this new basic knowledge about electromagnetism was soon applied to the generation of electricity. A general basis for applying it was supplied by John Hopkinson in a famous paper presented to the Royal Society in 1886. In collaboration with his brother Edward, John Hopkinson explained and systematized the fundamentals of electromagnetic circuits.[9]

John Hopkinson had been trained as a mathematician at Cambridge University and was employed as a consultant to the newly formed Electrical Construction Company, an Edison subsidiary, in 1882. Hopkinson redesigned the basic Edison dynamo so successfully that its output was increased by nearly 100 per cent. This new dynamo was manufactured by the engineering firm of Mather and Platt and quickly became a widely-used machine in England.[10] By the late 1880s dynamos incorporating new scientific knowledge were being manufactured. Up to this time the magnetic field of generators consisted of two poles, north and south. A new generation of multi-polar dynamos succeeded them, with four or more poles arranged alternately north and south. Multi-polar dynamos were more efficient, more powerful, and weighed less. They met the demand for increased current that came with the expansion of supply systems and the introduction of electric traction. The first successful designs were produced by American companies that had an interest in traction or large supply schemes. Thomson-Houston brought out a four-pole dynamo in 1890 which was rated at 100 kilowatts, and this was followed by the manufacture of a whole range of multi-polar dynamos.[11] In England, Gisbert Kapp designed a six-pole dynamo which was produced by W.H. Allen of York in great numbers. Such was the advance in dynamo design that a multi-polar dynamo which could exceed the kilowatt output of Edison's basic machine weighed one tenth as much.[12]

Although the dynamo is at the heart of the generating system there were also many component parts which were greatly improved during the 1880s. The drive between prime mover and dynamo was a source of breakdowns and voltage fluctuations which plagued the first electrical supply systems. Much of the "learning by doing" during the second phase was related to problems of drive linkage and voltage regulation, and a great deal of experience and expertise in this area was achieved by English engineers. The English marine engine builders, Willans and Robinson, designed a high speed prime mover which was directly coupled to the generator. This machine achieved world wide diffusion as the power source for supply systems.[13]

At the beginning of the 1880s some commentators on the electrical scene were convinced that electrical technology had reached a state of perfection. This was the conclusion of the president of the Society of Telegraph Engineers in 1880, who declared that "it is quite evident that there is no room for improvement left, even for inventors on the other side of the Atlantic."[14] This viewpoint is quite incorrect because the second phase of electrical development included the creation of an entirely new system of supply as well as the improvement of the existing direct current system. The alternating current system did not suffer from the disadvantage of small distribution areas which marred the development of direct current supply networks. The invention and introduction of alternating current technology was one of the most important developments of the second and third phases of electrical progress.

English engineers played an important role in the development of alternating current technology. There were many alternating current systems available in Europe during the second phase: Gaulard and Gibb, Zipernowski, Mordey, Kapp,

Westinghouse, Lowrie-Hall, Parson, Siemens, and Ferranti. Only three of these systems were of foreign design and manufacture, Siemens, Westinghouse, and Zipernowski. The first form of alternating current was one phase, so-called because the current flowed backwards and forwards in one wave. Its characteristics conform to a single sine wave of current flow which make it harder to understand (and to work with) than the non-alternating flow of direct current along a conductor. It is significant that an English engineer was recognized as the leading authority on the new current. Sebastian Ferranti's alternator designs were greeted with wide acclaim. He first produced a machine that had a very high output to weight ratio and followed this very successful design with a complete range of alternating current equipment. Ferranti's work as chief engineer to the Grosvenor Gallery Company marked the perfection of the single-phase alternating current system.[15]

The introduction of alternating current lighting systems was the first step in revolutionary changes in central station design which set the pattern for electrical development in the 1890s and 1900s. Sebastian Ferranti, whom Samuel Insull called the "father" of modern electricity supply, should be credited with the original conception of the modern power station.[16] In a strikingly prophetic analysis of the state of supply technology in the late 1880s, Ferranti pointed to the problems of direct current technology which would make it unsuitable for the future demands of public lighting. The central stations in city centers were designed around many small generating units and increases in demand were met by increases in the number of generators. This led to central stations, like the Elizabeth Street station in New York, which had 80 generating sets in 1889 with space provided for additions.[17] It was exceedingly difficult to bring large amounts of coal and water into central stations built in city centers, and it was equally troublesome and expensive to take away waste products. Furthermore, the noise, smoke, and vibration that accompanied generation were a great nuisance to city dwellers. In England, legal action was often taken to limit this nuisance, with the result that some stations had to be extensively rebuilt and others had to stop work until a solution to the problems of noise and dirt nuisance could be found.[18]

It was Ferranti's plan to take generation out of the cities and build central stations on rivers, or a communications hub, which would decrease transport costs and eliminate smoke and noise nuisance. Ferranti conceived of a very large generating unit which would produce electricity in bulk. As there were significant economies of scale in electricity generation, cheaper current could be produced and sold to a larger market. This plan depended on the development of technology of high voltage transmission to distribute current over a wide area and sell it to all classes of customers. The pioneer installation of this kind was made at Deptford, on the river Thames, just outside the center of London. Ferranti was made the chief engineer of the Deptford project which was built for the London Electricity Supply Company (Lesco). The plan was to use very large generators (rated at 10,000 horsepower) at Deptford and the current was to be transmitted at the unheard of pressure of 10,000 volts. The Deptford power station was to be the largest in the world and it was planned to supply all London and its suburbs, up to a total of 1,000,000 incandescent lamps.[19]

The Deptford project anticipated the development of supply technology of the 1890s. Unfortunately for Ferranti, Lesco, and the future of alternating current, the scheme was not a success. Its promoters admitted that it was "somewhat of an experiment" and the project suffered from the inevitable teething troubles, breakdowns, and accidents.[20] The abandonment of the project in 1892 marked the end of the great pioneering phase of alternating current technology. These first steps were not taken without cost. The failure of the Deptford scheme damaged the credibility of the technology involved and adversely affected its commercial future in England. Shops put candles in their windows to indicate that supply had ceased and accompanied them with notes reading "A Present from Deptford."[21]

The early start of English engineers in alternating current proved to be a disadvantage in the third period of electrical progress when more advanced multi-phase technology was being introduced. The Deptford fiasco deterred both investors and supply undertakings from adopting a technology in its experimental stage. Furthermore, the heavy investment in single-phase schemes led to a resistance to a technology which had more phases and different frequencies. Initial difficulties in making the two technologies compatible discouraged multi-phase extensions to single-phase installations. The concentration of engineering effort into single-phase technology diverted attention from multi-phase current. Ferranti and other English engineers spent a lot of time and money in the 1880s attempting to make an electric motor suitable for single-phase current. This proved to be a difficult task and there were very few commercially successful single-phase motors.[22]

The third phase of electrical progress involved the traction and multi-phase technologies in which England lagged. While the Deptford scheme was in its death throes, the Frankfurt-Lauffen transmission was proving that high voltage distribution was viable. The management of the company that had undertaken much of the work on this transmission had originally been hesitant about committing its resources to an untried technology. Emil Rathenau, who headed AEG, was informed of Ferranti's work at Deptford and was critical of risks being taken and of the experimental nature of the project. He delayed the introduction of the new technology until he was sure that all development problems had been solved. Rathenau's caution was well repaid, for the Frankfurt-Lauffen transmission began a wave of diffusion of the new technology in which AEG played a leading part.[23] The successful exhibition at Frankfurt assured the commercial acceptance of the technology, and the 1890s became the decade of multi-phase supply systems.

The popularity of multi-phase systems in the United States and the Continent increased during the third phase and so did criticisms of England's lag in this technology. It was noticeable that the foreign engineers who developed this technology had the benefit of advanced scientific and engineering training. Likewise, observers of the electrical industry could see that the large foreign companies dominated the market for this advanced technology. It was against this background that the image of backward electrical engineering in England was fostered. Yet the development and marketing of the new technology, especially the highly successful three-phase systems, indicate that scientifically trained personnel were not the only advantage possessed by foreign electrical manufacturers.

The most important part of the three-phase system was the induction motor. This machine was based on the principle of the rotating magnetic field and solved all the problems which plagued alternating current motors. The discovery of the rotating magnetic field is credited to Nikola Tesla, who had attended the Technische Hochschule in Gratz, Austria. He maintained that he built an induction motor employing multi-phase current as early as 1883.[24] Professor Galileo Ferraris, of the University of

Turin, also discovered the properties of the rotating magnetic field and built a small induction motor in his laboratory in 1885. Yet the discovery of the induction motor was only the first step in a long process of research and development in which scientists and inventors did not play the leading role. Neither Tesla nor Ferraris were able to promote their discoveries alone; in fact, Ferraris did not appreciate the commercial importance of his findings and did not bother to publish them until 1888.[25] Although Tesla was employed by the Edison companies in both France and the United States, he failed to interest his employers in his inventions, and by 1886 he was reduced to working as a day laborer.[26]

The key to the innovation and development of multi-phase technologies lay in the resources of a large company with interests in both electrical manufacture and supply. Three-phase technology can not be adopted in a piecemeal fashion because it must be incorporated into a complete system which is compatible with currents in use. Therefore, converters had to be designed which would enable users to convert three-phase into whatever current was needed to operate their machines. Transformers had to be built to handle the new high voltages and there was also a need for new switches and cables. Furthermore, the laboratory induction motor had to be perfected into a viable commercial product that would be the cornerstone of the new system. A project of this magnitude needed a great amount of finance and marketing skills as well as engineering expertise.

The experience of the Westinghouse Company in developing its multi-phase system illustrates the point. After acquiring the Tesla patents in 1888, the Company devoted large amounts of time, human capital, and money to the project. This work went on for seven years before the system was put on the market. It involved the attention of many of Westinghouse's best engineers and was a factor in a severe financial crisis which had to be overcome before the project could be completed.[27] Marketing the new technology posed more problems. The Westinghouse Company correctly perceived that industrial users were satisfied with direct current motors and would be unreceptive to the new current. It was therefore decided to sell multi-phase technology to supply systems in order to build up demand for equipment to suit the new current.[28] Pushing the system in the supply industry involved a great deal of marketing expertise and financial assistance. No company in England in the third phase could successfully manipulate the supply industry in this manner.

There was no English electrical manufacturer who could match Westinghouse's resources in developing and marketing the multi-phase system. The best vehicle for the diffusion of this technology was the multinational company with extensive engineering and financial resources. The advantages of such an organization can be seen in the development of the electric tramway: the other area of English technological lag in the third phase. The technology of electric traction involved little applied science because the greatest problems were as much mechanical as electrical in nature. The essence of traction technology was "learning by doing," for traction schemes involved many component parts that had to work together under difficult conditions. For example, the development of traction motors was governed by experience accumulated in the operation of tramways. The problems of speed regulation, overheating, and making the equipment resistant to dirt and vibration could only be solved after engineers had carefully observed the day-to-day operation of a traction system. This kind of product development could only be carried out by a large staff of engineers who had experience in both mechanical and electrical components of traction technology. In the manufacture and design of traction equipment, there was no substitute for experience. The large American manufacturers that dominated the traction market made a policy of acquiring experienced personnel as well as keeping in close touch with traction operators. The information and personnel gathered in this manner increased the store of "learning by doing" of the manufacturers.

English electrical manufacturers had little opportunity to gain experience in traction technology. When the municipalities entered the traction market they had a choice of a tried and tested American system or an experimental English system. The municipalities naturally opted for complete systems that were well past the experimental stage. This was a wise choice, but it did not give the English industry a chance to acquire the necessary experience. This situation also existed in the market for three-phase systems. The predominant system in use in the English supply industry was direct current because this best suited the small station and distribution areas that characterized the average undertaking. The municipal opposition to large distribution areas prevented the formation of large supply networks and also denied English manufacturers a chance to gain experience in this technology. When the large municipal undertakings decided to expand their systems and thus entered the market for three-phase equipment, they chose equipment offered by experienced foreign manufacturers rather than the first efforts of English electrical manufacturers. This gave English engineers little opportunity to gain experience in three-phase technology.[29]

There was no lack of highly trained personnel in the English electrical industry; what was lacking was the experience in electrical technologies which involved heavy engineering and very large power systems. Cardwell has made the point that many trained engineers came to England to market their education and skills.[30] Some of the best electrical engineers in the world chose to work in England and made immeasurable contributions to the development of English electrical technology. Gisbert Kapp, for example, came to England to work for Crompton, then returned to Germany to hold the post of general secretary of the Verband Deutscher Elektrotechniker. He spent the rest of his career in England and finally occupied the chair of electrical engineering at the University of Birmingham.[31]

Chapters 1 and 3 described the agitation for improved technical education in England and noted the continual comparison with the highly vaunted German system of technical education. Although the English system of technical education was the target of much criticism and unfavorable comparison, it did turn out a steady stream of engineers who entered the electrical industry. Shadwell has made the point that English technical education in the nineteenth century was unjustly depreciated because its great progress in the last half of the century was not given sufficient credit.[32] While the Germans trained their engineers from above, in Technische Hochschulen and universities, Shadwell argues that the English trained from below, in night schools and extension courses. Although the English method of technical education was not as systematic and rigorous as the German, Shadwell notes the great progress achieved after the government entered the field in 1853.[33]

Shadwell's conclusions are supported by Roderick and Stephens who argue that English society in the latter half of the nineteenth century "became increasingly professional... in which the trained scientist and technologist replaced the amateur."[34] While formalized technical training in England was neglected, there were an increasing number of ways in which an individual could acquire a technical education. As Roderick and Stephens note, Victorian England was committed to the creed of self-help.[35] Part-time education in electrical engineering was begun in 1879. Night classes and part-time courses were soon supplemented by full-time courses in the new provincial technical colleges, such as Owens College in Manchester. In the 1890s the newly founded redbrick university colleges offered courses in electrical engineering.[36] Some of the leading figures in the English electrical industry had acquired technical education in night schools or part-time courses, including Sebastian Ferranti, William Mordey, Thomas Parker (of Elwell-Parker Manufacturing Co.), Oliver Lodge, and the consulting engineer John Aylmer.[37]

In addition to technical training, there were several ways in which an English engineer could acquire practical experience in the basics of electrical engineering. The multinational companies provided excellent training in advanced electrical technology as well as giving instruction on its manufacture and marketing. The Siemens Company in England was credited with an excellent training program, and Siemens-trained engineers were sought after in the job market.[38] It was also customary for foreign companies to send English engineers to the home base for training in advanced technology. The Westinghouse Company, for example, sent large groups of young men to the main works at Pittsburgh. The "Holy Forty," as one group was called, were later to make their mark as leading figures in the English electrical industry.[39]

There was no lack of trained personnel available to the English industry in the third and fourth phases of electrical development. On the contrary, there appears to be a surplus of engineers on the job market during this period. The electrical press often commented on the competition for jobs in the industry, the low value given by employers to highly trained engineers, and the low wages and poor conditions in municipal supply undertakings that preferred to employ the untrained and unskilled.[40] Low wages and lack of demand for trained engineers shifted attention to the demand side of technical education, and it appears that English industrialists did little to stimulate demand. The blaming of the lag on poor technical education suited the arguments of those campaigning to increase financial support of technical education. There can be little doubt that many of the calls for better technical education were, in fact, "vague cries" that sprang from a misguided belief that technical education was the panacea for all industrial ills.[41]

There appeared to be few barriers to the diffusion of electrical information to England in the third and fourth phases. The archives of English electrical manufacturers, supply organizations, and engineering firms are full of information about new technology. The grand tour of foreign installations was commonplace in the English industry and a great deal of time and expense was incurred in arranging visits and meetings. The dissemination of technological information was carried out internationally and English engineers and entrepreneurs were widely travelled. There were few problems in acquiring new technology--what was lacking in England was the demand for it. Even if there had been a lack of trained engineers in England, the multinational companies would have been glad to provide them. The chief problem these firms faced in England was obtaining the business for traction and multi-phase schemes.

The large three-phase and traction schemes were the most visible new technologies of the third phase. There were other, less visible, technologies which were equally important. These technological advances were not in the public eye and were therefore not factors in the perceived lag. The improvements in cables, switching, and controlling provided the foundations for electrical progress in the 1890s. For example, the introduction of the paper covered cable was a vital part of new electrical systems and its development and introduction owed much to English engineers and cable manufacturers.[42] These developments were not mentioned when the lag in electrical technology was being discussed in the popular press.

The heritage of Ferranti was evident in the third phase because English engineers made great strides in the creation and operation of large supply systems based on "super" power stations. Charles Merz was recognized as the leading central station engineer and his designs had great impact on the technology of electricity supply in the twentieth century. Merz's first important central station was built at Neptune Bank in 1899, and it was the result of the great increase in demand for Nesco's light and power. The new station was the first to incorporate the steam turbine, which was lighter, more powerful, and took up less space than existing prime movers. The introduction of the steam turbine began a revolution in power station technology because it made possible an enormous increase in output at much lower costs.

The successful Neptune Bank station was followed by an even larger station built at Carville in 1902. These two "super" power stations were milestones in power station engineering and were widely copied. Merz became a consulting engineer with an international reputation. His work in India and Australia was crucial to electrical development in these countries, and he also played an influential role in the extension and enlargement of the American supply industry. The design of the Forth Street Power Station in Chicago was an important innovation in American supply technology and one of the most influential projects of Merz's career. This station, and the stations which were based upon its basic design, played an important part in determining the development of power station technology in the United States.[43]

The "super" power station became one of the most visible

aspects of the technology of the fourth phase. The creation of very large supply systems with an extensive distribution network was the first step towards the formation of a national grid. Much of the new technology of running this system was in the form of "learning by doing," for the experience gained in managing large distribution networks was essential in the development of our modern system of supply. Many of these lessons were learned in the Nesco supply undertaking, which expanded very rapidly in the fourth phase. The joining of Nesco and two other bulk supply networks formed the largest supply system in Europe and provided the basis for the English grid.[44] The work of Charles Merz in the Nesco undertaking was to provide the foundation for the general expansion of English supply schemes during the fourth phase.

Much of the engineering effort in the fourth phase went into the development of high voltage alternating current technology. The large foreign electrical manufacturers had invested a great deal of capital in this technology, and continued to use their considerable resources to improve their multi-phase equipment. The research into high voltage technology was primarily aimed at reducing the cost and increasing the reliability of the component parts. Much of this work was carried out in specialized departments which were the forerunners of research and development units. Although some leading scientists were employed in these departments, the goal of organized research was oriented towards problem solving in existing technologies rather than the creation of new scientific knowledge. The general laws of electromagnetism were still those formulated by Maxwell in the nineteenth century. As Reich comments, "To a manager concerned with meeting the competition in matters of service, price... and technical capabilities, research meant working towards straight forward technological goals."[45]

The multinational companies used their research departments to improve their hold on some of the important parts of three-phase technology, such as alternators, transformers, and rotary converters.[46] In this way they were able to dominate the market for these items during the twentieth century. There were other parts of the system that were not in the hands of the multinationals. English engineers and electrical manufacturers played their part in the improvement of switching, controlling, and regulating gear. British expertise in boiler and steam pipe technology were highly regarded in the American electrical industry and many installations bought, licensed, or copied English technology.[47]

Although the research departments of the multinationals gave the appearance of technological superiority to contemporaries, and to some modern commentators, the large foreign companies were technological leaders only when it suited them. Their development of the key technologies of the third phase gave them a lead in traction and three-phase equipment which they marketed aggressively in the 1890s. Yet their great interest in the technology of the third phase made the German and American multinationals loath to make it obsolete by further innovation. For example, the American companies with interests in direct current traction resisted the introduction of three-phase traction schemes and were heavily criticized in the American electrical press for being technologically conservative.[48]

There was no noticeable technological lag in England during the fourth phase, and the excellent electrical engineering of this phase was built on solid foundations. Byatt has argued that English electrical engineers performed much better in the 1880s than in the 1890s.[49] Yet the technologies of three-phase current and traction were developed in the late 1880s. The new technologies of the third phase were built on the experience and expertise acquired in the previous phase. The long gestation period of the induction motor has been described, and it appears that the important early research into high voltage currents was made in the late 1880s.[50] Similarly, the improvements in power station engineering, which dominated the technology of the fourth phase, were the result of the engineering work of the late 1890s.

Byatt has underrated the quality of English electrical engineering of the 1890s. The lags in traction and three-phase current were important, but to interpret them as signs of technological backwardness is to overlook many engineering developments of the third phase. The development of the English electrical industry was governed more by institutional than engineering factors. It was both unfair and incorrect to place the blame for slow development of electrical supply onto the engineering fraternity, for as Ferranti argued in the Institution of Electrical Engineers, the system was not of their making.[51]

English electrical engineers were highly regarded and recruited by multinational companies. Some of them rose through the corporate ranks to be the leading engineers of their time. For example, Francis Hodgkinson became "the Dean of the steam turbine" in the United States after Westinghouse had lured him from Parsons.[52] With English electrical engineering represented by men like Hodgkinson, Merz and Ferranti, the late Victorians' concern with engineering backwardness was misplaced. The Weir Report of 1916 singled out Samuel Insull as "the greatest expert on electrical problems in the world"[53] Insull's account of electrical development was based on the pioneering work of three great electrical engineers--all of them English. In a speech given before some of the most important figures in the American electrical industry, Insull argued that Sebastian Ferranti, Charles Merz, and Arthur Wright (of the Brighton undertaking) had provided the basic elements of the modern system of electricity supply. Of Merz, Insull said: "those companies which are engaged in selling electricity on a large scale, owe more to the work of Merz than to any other man."[54]

CHAPTER SEVEN

THE FORMATION OF MARKETS FOR ELECTRICITY

The pioneers of electric lighting faced numerous problems in the creation of viable lighting systems, and not the least of which was convincing a wary public of the safety and advantages of the new light. The first electrical engineers thought that once the advantages of the new illuminant were publicized there would be few problems in selling it, but their expectations proved to be over-optimistic. Selling electricity in its many forms was not easy for the marketing of the new light involved the creation of sales technologies which took many years to perfect. This involved, in turn, a new level of cooperation between engineering and marketing.[1]

The primary market for the electric light was the existing market for gas light. The first electrical systems builders, men like R.B. Crompton in England, Johann Schuckert in Germany, and Thomas Edison in the United States, saw electricity as a replacement for gas. Furthermore, the existing systems of gas distribution influenced the development of its new competitor. For example, Edison's idea of the central station owed much to the gas works supplying many customers from one central plant.

As direct current flowed along conductors like gas along a pipe, it was a simple task to copy the system of gas distribution, replacing gas with electricity. Edison spent some time studying books on gas illumination before embarking on the planning of his first central station.[2]

The marketing of gas provided the main example to follow in the first efforts to market electricity. Yet this method was primitive in the 1880s. Most gas companies normally sold lighting for a fixed period of time each day. For a yearly or monthly rate, gas users were allowed to burn lights between certain hours, and after the "lights out" time the gas companies employed men to check that all gas burners were extinguished. This system depended on the honesty of the customer and many of them were annoyed at the call of the gasman to turn off their lights after the designated time. The majority of the first electric lighting contracts copied the gas industry's system of pricing. Therefore the first users of the electric lighting paid a fixed sum for the use of a specified number of lamps during designated time periods. Rates were normally paid by the month for use of arc or incandescent lamps, but street lighting contracts were in force for a year and stipulated exactly when the lamps were to be illuminated and extinguished.

The selling of electricity for a fixed rate per lamp was highly disadvantageous for the producers of electricity. They knew little of the daily demand for lighting and practically nothing about seasonal variations. It is clear from the evidence given before official bodies and from the electrical press that there was little knowledge about the costs of generation. The pioneers of electric lighting were ignorant of patterns of demand and could only make vague guesses about their costs; consequently their attempts to replace gas light met with little success. Many claims were made about the greater economy of the new illuminant, but as the Journal of Gas Lighting pointed out, these claims were based on optimism and guesswork and had little foundation in fact.[4]

One of the major problems in the marketing of electricity was the estimation of consumption. The development of meters was therefore a very important step forward in the second phase of electrical development. The early meters were primitive and inaccurate, but they did establish a framework in which to estimate electricity consumption. The first meters in commercial use were the Edison meters invented for the Pearl Street distribution system. Like many other components in this system, the meter was designed from scratch. Edison chose to exploit electrochemical phenomena in the design of his meter, which consisted of two sheets of zinc immersed in zinc sulphate. The passage of current transferred zinc from the positive to negative sheet. The sheets were then weighed and the difference in weight provided the basis of calculations to find out how much current had been used. This method was expensive, time consuming, and highly inaccurate.[5] The Pearl Street operation provides us with the first examples of disputes over electricity bills. This situation was repeated wherever electricity was introduced as a replacement for gas lighting.

A considerable improvement in meters occurred when new knowledge about magnetism was applied in the third phase of development. Advances were made in the manufacture of small springs and in magnetic materials with the result that a meter employing magnetism to measure electricity was possible in the 1890s. After the Paris Exhibition had failed to produce a commercial meter in 1889, a prize was offered to the inventor of an improved meter. Elihu Thomson won the prize and his design became the starting point of the improved meters brought out in the third phase.[6]

The crude marketing of electricity in the second phase brought little competition to the gas industry in England. Although the supporters of electricity stressed its cleanliness and flexibility, the main selling point of the new illuminant was price. In England, Germany, and the United States, the pioneers of electric light thought they could produce a better and cheaper light. They were all to find out that electric lighting was a superior but more expensive light. The appeal of the electric light directly corresponded to the price of gas, and where gas was expensive, adoption of electricity was fast and widespread. On the other hand, where gas was cheap, electricity was slow to be adopted.

The price of gas determined the size of the market for electricity and the great difference in gas prices explains the differences in regional patterns of adoption. In the second phase of electrical development the price of gas in the United States ranged from 8s to 12s per 1000 cubic feet. This was much more expensive than the price of gas in England. The average price per 1000 cubic feet was between 2s 6d and 3s, compared with the American average of around 10s. German gas prices ranged from 7s per 1000 feet in Bremen to 4s 6d in Berlin. Average prices were around 5s to 6s, with lower prices prevailing in the Ruhr industrial region.[7]

The cheapness of gas in England deterred potential adopters of electricity, and institutional factors made English electricity more expensive, on average, than electricity generated in Germany and the United States during the second phase. In addition to wiring regulations, there were provisions for compulsory supply which added to the cost of electricity. Electrical undertakings in the United States and Germany were free to supply whomever they wanted. In this way pioneers like Edison were able to choose the best customers for his Pearl Street station. Decisions to connect up a customer's premises were made on the estimated amount of demand and on the times that electricity was needed. It was better to arrange a steady demand from a concentrated group of customers than a demand which had evening and morning peaks. The peak demand determined the generating capacity that had to be installed in central stations. A high peak demand meant that a great amount of generating equipment was not used during periods of slack demand; this increased costs and was reflected in the price of current. The provisions of the Electric Lighting Act of 1882 made it difficult to arrange a favorable schedule of lighting demand in England, because the undertaking was forced to supply any customer within fifty yards of the main.[8] This denied the supply undertaking the flexibility needed for economical operation of its generating plant.

Several other provisions of the Act had direct bearing on the price of electricity. There were heavy penalties on the interruption of supply, and as the primitive state of electrical technology caused many breakdowns, the English undertaking was often made to pay the price for inconstant supply. The Act also fixed the maximum price for electricity which curtailed the activities of supply undertakings in the luxury market for electricity. The high relative cost of electricity during the second phase priced it out of the market for domestic and public lighting in England. The 1882 Act placed the supply industry into the hands of the municipalities and they were unwilling to adopt a technology which was much more costly than the existing illuminant. The English gas companies responded to the challenge of electricity by cutting their own prices and giving their customers better service. The experience of this industry proves that "threatened men live long" for the increased value of its byproducts enabled the gas industry to reduce its costs and prices up to the beginning of the twentieth century.[9]

This set of institutional, technological, and economic factors forced the English electrical industry into a very narrow market for its service. The higher relative costs of the new illuminant led the electrical undertakings to market their product on the basis of its convenience, attractiveness, and novelty value. The first customers for the new light were often found in the retail trade and in the service sector. Hotels, theaters, clubs, and other places of entertainment were quick to adopt the new light. Electric lighting was cleaner and did not give off obnoxious gases, nor did it raise the temperature in confined areas. Some customers of the electric light claimed that its increased cost was justified by the reduction of cleaning costs. The annual redecoration of shops and hotels was partly the consequence of using gas light which blackened walls and ceilings.[10]

When it was first introduced, the electric light had a novelty value which endeared it to the owners of shops, hotels, and theaters. The ease of switching and varying the light also made possible the use of lighting for dramatic effect. Large retail operations were quick to appreciate the potential of the new light. It could be used for display and promotion, as well as for lighting. In the United States it was soon said that "all good stores must have the electric light."[11] In England, Germany, and the United States, the largest installations of electric lights were in theaters.[12]

The next market to be conquered was the market for public lighting and the industrial lighting of large mills, workshops, and other areas which needed flood lighting around the clock. English entrepreneurs found these markets harder to penetrate than their American and German equivalents. The series of public exhibitions which preceded the public adoption of street and building lighting failed to prove that electricity was cheaper than gas lighting in England. The failure of the experimental installations to provide a cheaper and more constant supply of light substantiated the claims of the gas industry. It was obvious that the promoters of electricity had an untried and expensive technology which could not live up to their exaggerated claims. Although many of these experimental installations were made below cost they still could not compete with gas in terms of price.[13]

Textile mills, printing works, and railway yards were all quick to adopt electricity in the form of the arc lamp which could be used to light very large areas. Other industrial users took longer to innovate and needed more persuasion. In Germany and the United States the industrial user was courted by the electrical industry from the earliest introduction of lighting.

The availability of hydroelectric power was a great inducement to adopt electricity because it lowered the price of current and gave the user the means to control his power source. The German and American electrical manufacturers immediately recognised the potential of industrial use of electricity and developed products for it. The Siemens and Halske Company conceived its traction schemes entirely within the framework of industrial use and also developed a line of motorized equipment and lighting that was designed for industry.[14] The efforts of the electrical manufacturers in the United States were concentrated in the industrial northeast. Spokesmen for the industry estimated in 1886 that the majority of New England's textile mills were lit by electricity and that roughly 90,000 incandescent and 6,000 arc lamps were in use.[15]

The English electrical manufacturers did not have this orientation towards the industrial user because their attention was on the domestic and commercial user. The commercial and entertainment centers formed the nucleus of demand for electricity in England and subsequently the luxury market became the main market for electricity in the 1880s. The electric light had to be marketed as the light of luxury because it could not compete with gas in any other market. R.B. Crompton, one of the first English electrical manufacturers, saw that this market would be crucial to the development of the electrical industry in the 1880s and concentrated his efforts to supply small wealthy neighborhoods and the stately homes of the upper class. Crompton believed that installations in such prestigious surroundings would bring good publicity for the new illuminant and he realized that cost was a secondary consideration to the customer.[16]

The formation of the English supply industry centered around supply schemes to resort areas like Brighton and Folkestone, and also to fashionable London suburbs. The first successful lighting undertakings in England were at Brighton, Kensington, and Knightsbridge. These first installations had several similar features. They were normally small undertakings that had been formed or financed by groups of wealthy individuals who wanted the new light in their homes. In many cases the center of the distribution area was a block with a high concentration of hotels, theaters, and other places of entertainment. The Brighton undertaking was centered on a seafront block of hotels, arcades, shops, and restaurants. The concentration of private property at the center of the supply area made it easier to avoid wiring and other costly regulations. For example, the exclusive Kensington Court residential estate was supplied by cables running through its underground conduits, and the Brighton company strung its wires (illegally) from the tops of the hotels it served.[17]

Although development of the luxury market did not provide the spectacular evidence that accompanied foreign development of street lighting, it did provide business for a small but prosperous electrical manufacturing industry. The Crompton Company did excellent business in isolated domestic lighting, supplying equipment for private lighting undertakings and ship lighting in the 1880s. Business was so good that the company instituted a night shift in the late 1880s.[18] This prosperity did not prevent the company's founder from commenting on the slump of the English electrical industry in the debates of the Institution of Electrical Engineers and in the press. The American electrical industry was aware of the good business and high profits enjoyed in the English luxury market, and comments in the electrical press indicate that American manufacturers believed that this was an excellent market to exploit.[18]

The unique market orientation of the English electrical industry influenced the development of the electrical manufacturers who remained small in size and limited in the scope of goods and services they provided. On the other hand, the exploitation of the market for public supply and industrial use gave the impetus for the formation of large companies (in the United States and Germany) which were able to offer a full range of electrical services. Until the municipalities in England commenced public supply, the English electrical manufacturers had little opportunity to engage in heavy electrical engineering such as the construction of large power stations and long distribution networks. Their business was in small isolated plant and therefore they remained small concerns. The great amount of capital needed for heavy electrical engineering placed a premium on finance in the electrical industry. While the German and American firms formed close relationships with large banking and investment concerns, their English colleagues found it extremely difficult to obtain finance. The English companies constantly complained about the difficulties of raising capital.[20] The electrical bubble of 1882 had given electrical risks a very bad name in the London money market and this stigma remained throughout the 1880s and 1890s.[21]

The English electrical manufacturers had a very narrow outlook and tended to use their local connections to obtain business. Very few attempts were made to penetrate the Continental market for electrical goods. Only one company, Crompton, made a large installation on the Continent. The Crompton lighting system in Vienna was a great success and could have opened the door to an invasion of the lucrative electrical market of Central Europe and Italy. The Crompton company found it impossible to obtain the finance needed to compete with German companies in the central station market, and this lack of capital made it difficult to match the German companies' financing plans which accompanied their tenders.[22] Unlike the English municipalities, the local authorities on the Continent made sure the business went to local companies.[23] This was an important factor in the reduction of the potential market for English electrical goods. The protectionist policies of Continental and

American governments in the 1890s also served to close their electrical markets to English companies.

The municipal lighting boom of the third phase of electrical development gave English manufacturers an opportunity to enlarge their operations and diversify their products. Unfortunately the municipalities chose to use a system of consulting engineers and tendering which hindered the emulation of the American and German industries. In contrast to the municipal authorities in Germany and the United States, English municipalities entrusted the planning and design of their installations to independent engineers. The practice on the Continent and in the United States was for the contracting electrical manufacturer to provide the design and engineering expertise in the building of the entire system. The English consulting engineer would draw up specifications and then manufacturers were invited to tender for the component parts of the system. The use of consulting engineers was criticized in the electrical press and in the I.E.E. It was claimed that the use of consulting engineers encouraged non-standardized plant and severely weakened the power of the manufacturer in the market place.[24] It was very difficult for one company to install all parts of a lighting system. When the use of consulting engineers was extended to traction schemes the results were similar, for no single company was allowed to build complete systems. This practice did little to encourage the increase in size of electrical manufacturers. In contrast, the marketing and sales practices in German and American markets aided the formation of very large electrical manufacturers.

The optimism at the beginning of the era of electric lighting encouraged the establishment of very many electrical firms in the first years of the 1880s. The bursting of the electrical "bubble" in 1882 put an end to the majority of the English electrical companies. With public lighting firmly in the hands of the local authorities, there was little point in forming private companies for lighting.[25] The great number of small companies in Germany and the United States produced an environment of intense competition in which a strong marketing department was a prerequisite for survival. Rather than follow the English example of applying to the government to light a specific area, the American system was one of franchise and close cooperation between electrical manufacturer and supply company.

The first American electrical manufacturers enlisted the help of local businessmen and politicians in the formation of supply undertakings. The electrical manufacturer accepted the stock of the franchised local company in return for lighting equipment. In this way a close bond was forged between manufacturer and supply undertaking which assured the former of a steady demand for its goods and services. The key to American electrical development was the union of interest between manufacturer and supplier. The Edison company was a pioneer in this practice because Edison believed that the manufacturer should have the faith in the quality of his equipment and in the future of public lighting. The Edison company kept a strong financial interest in its local affiliates and sold them equipment at factory prices in return for stock and a share in net earnings. The parent company also provided free advice and technical assistance to its franchised supply companies.[26]

Increasing competition in the lighting market led to more advanced forms of financial arrangements between electrical manufacturers and the supply undertakings. The manufacturers formed companies to underwrite the new lighting ventures and obtain the great amounts of capital needed. The Thomson-Houston company formed a subsidiary which took over the stocks of local utility companies and used them as collateral to sell debentures to the public. The Edison interests organized the Edison Construction Company which provided the credit and financial assistance to the franchised utilities. Samuel Insull was put in charge of this operation and his financial expertise was to play an important part in the spread of the Edison lighting system.

The German electrical manufacturers followed the American practice of forming and financing the supply companies which provided the basis of demand for their products. Initially the German electrical companies emulated the English practice of non-involvement in the supply industry. Werner Siemens had originally insisted that his company should not get involved in the operation of supply companies. In 1893 he made an agreement with Emil Rathenau, who had acquired the Edison patents and formed a company to exploit them. The German Edison Company (DEG) retained the right to manufacture the Edison incandescent lamp, but the terms of the 1893 agreement gave Siemens the sole right to manufacture equipment for Edison central stations. The other provisions of this agreement gave Siemens and DEG a monopoly of manufacturing Edison equipment and forced the other German manufacturers like Schuckert to buy the lamps they were busy installing in Central Europe.[28]

The DEG did not keep this agreement for very long. As soon as it was obvious that the Pearl Street station was a commercial success, the DEG negotiated with the Berlin local government to build a copy of the Pearl Street station in Berlin. The station on the Markgrafenstrasse was to become pivotal to German electrical development for it provided the model for future development. One of the most important examples to be followed by the rest of the industry was the method of financing the station. The DEG agreed to pay for all the costs of obtaining the franchise and building the station. A separate company was formed to run the system and manage its expansion. In return for this massive outlay of capital, the DEG obtained the sole right to supply all the equipment needs of the system. Althought the financial arrangements of the Berlin Electricity Works (BEW) were not made public, they provided the German electrical industry with an example of creating a market for electrical goods that was copied on a large scale.

The DEG had taken a great risk in underwriting the BEW, but as demand for electric light in Berlin proved to be insatiable, both organizations prospered. By 1887 there were seven central stations in Berlin supplying 252 arc and 17,800 incandescent lamps. In the same year AEG was formed to take over the DEG's interests and to provide a wide range of electrical goods. AEG became the main competitor of Siemens and Halske in the manufacture of lighting equipment.[30]

The great success of the BEW encouraged the formation of many more companies to exploit the boom in public lighting in Germany. With local companies taking much of the trade in isolated plants, the larger electrical manufacturers began to concentrate on the financing and building of large central stations. In much the same way as the American electrical industry, a small number of very large electrical manufacturers came to dominate the entire industry. The development of electrical technology in the third phase accelerated this process, for large traction and lighting schemes needed the large-scale technological and financial capabilities that were only in the hands of the big companies.

As Chandler noted in <u>The Visible Hand</u>, the traction schemes demanded very large organizations both financially and operationally. The 1890s became the age of consolidation in the American and German electrical industries. The owners of the new organizations became adept at raising funds, forming consolidations, and doing the "necessary political wheeling and dealing."[31] The intense competition in the electrical manufacturing industry favored the formation of large enterprises which devoured the smaller companies. The manufacturers increased their financial support of the supply and traction companies which followed the pattern established in the 1880s.

The practice of creating demand for electricity by financing supply organizations was not without pitfalls, and some of the disadvantages of this practice became obvious in the 1890s. "Customers did not exist; they had to be created"[32] was one conclusion on this process, and this inevitably led to undertakings that were incapable of making a profit because demand was not high enough to justify expenses. It was noted in the American press that many undertakings were more the result of the competition between manufacturing companies than the answering of the consumer's needs.[33] Many of the supply companies had short and unprofitable lives and the successful stations of the third phase often stood on the remains of several unsuccessful stations of the second period.[34] While American electricians were arguing that many undertakings had been established ahead of demand, their English colleagues were complaining about the opposite. In England many communities who wanted the electric light did not get it because their local governments would not permit a private company to commence supply.

The difference in marketing practices can help explain the dissimilar paths of development and styles of organization found in the countries under discussion. The accelerated growth of the German and American electrical industries was based on a rapid expansion of supply that was financed by the manufacturers. Yet the great competition among manufacturers meant that many unprofitable supply undertakings were established. The large number of central stations that so impressed visiting engineers from England included many that were destined for bankruptcy. The failure of local supply companies reached a climax in the depressed years of the 1890s and almost ruined the large electrical manufacturers who held their stock. By 1890 the Edison interests held about $4,000,000 worth of stock in supply companies and the financial strain of bearing this burden led to the formation of General Electric in 1892. The new company soon found itself in a worse position. When a financial panic occurred in 1893, GE was holding over $16,000,000 worth of local companies' stock. Most of the stock was worthless and the best of it had to be heavily discounted.[36]

The German industry followed the same process. There was the same intense competition and forced growth of the supply industry. None of the many new supply undertakings were as profitable as the BEW and consequently many failed. The manufacturers who held their stock also got into financial difficulties and many were taken over by the largest concerns. By the end of the 1890s the great boom in central station and traction schemes came to an end and only the strongest manufacturers remained. Competition intensified for the remaining business and both American and German electrical giants saw their salvation in export markets. The Germans widened their scope and sought new business in Scandinavia and southern Europe.[37] The Americans aggressively marketed their goods all over Europe and alarmed the Germans who were used to a dominant position in the European market.[38]

England became an important export market for German and American electrical concerns and this led to the "invasions" that marked electrical development by the turn of the century. These inroads into the British market brought a wave of criticism of the home industry and increased perceptions of a lag. The entrance of the foreign multinationals during the third phase of electrical development naturally focused attention on the smaller, and financially weaker, English electrical manufacturers.

The entrance of foreign companies in the late 1890s came at a time of unprecedented activity for the English electrical manufacturers. The beginning of the traction boom had coincided with a second wave of central station building, and the English companies were fully stretched to meet this increased demand.[41] The orders that went to foreign firms were usually contracts that

English companies were unable to meet, either because they were too busy or because the projects were too big. The German and American multinationals arrived in time to take part in a traction boom that came several years after similar booms in their home markets. They did not take business from the English manufacturers and leave them ruined, as was often depicted in the press.

The large contracts that went to the multinationals was often interpreted as an indictment of the backwardness of English electrical engineering and the weakness of the home industry. In reality, they proved to be valuable to the development of the English electrical industry. English companies were making good profits and learning from the multinationals in their midst. In the fourth phase the English manufacturers were able to put into practice many of the lessons they learned in the previous phase. Companies began to copy American manufacturing techniques and German sales organizations. Some new technologies were diffused by foreign companies at this time, and among them were some important new ideas in marketing and sales.

CHAPTER EIGHT

THE MARKETING OF ELECTRIC POWER

The marketing of electrical goods and services played an important role in the diffusion of electrical technology. The primitive methods of selling electricity to the public during the first and second phases were instrumental in the slow diffusion of electric lighting. Little was known about pricing the current and measuring its use by the consumer. Great progress was made in marketing and sales during the third phase. The system of fixed rates and price competition with gas were recognized as unsatisfactory methods of promoting the use of electric light. The Edison interests and the American Institute of Electrical Engineers soon realized that a "cardinal mistake" had been made in their marketing.[1] New methods and new technologies were developed in the third phase, ranging from the introduction of better meters to a complete restructuring of tariffs.

The main problem in selling electricity at low prices was the construction of a favorable load factor in generation. The load factor of a power station is the ratio of actual output to the output that would have resulted if the station had operated continuously at full capacity during the measurement period. The load factor is an indication of capital utilization and also governs the price at which electricity can be sold. The load factors of the second phase were very poor because central stations were built to supply electricity for lighting only. The National Electric Light Association of the United States noted that in the early years of electric lighting many central stations sold from ten to thirty per cent of their generating capacity, being idle about eighty per cent of the time.[2]

A typical lighting load at this time had evening and morning peaks. The high peak load in the evening was the only time that central stations operated at maximum output.[3] Load factors that were solely based on lighting were responsible for maintaining high costs in the form of excessive reserve capacity that was used in peak periods. The acquisition of a traction load in the third phase did little to improve load factors because peak traction demand tended to coincide with peak lighting demand.

The improvement of load factors came as a result of supply organizations exploring the market and actively courting customers with favorable consumption patterns. As long as the price of electricity was based on the amount of current used, there was little opportunity for the manipulation of demand that was essential for a good load factor. The consumer who needed a unit of electricity at 6 p.m. cost the generating station more than the consumer who needed a hundred units at 12 p.m. The answer to this problem was a restructuring of rates to reflect the costs of generation. The maximum demand system became the basic method of charging the customer the costs of supplying him. This system based rates on the maximum demand, and the higher the demand, the lower the rates for each additional unit. Thus customers using current for the longest period during the day obtained the lowest rates. In addition, peak hour rates were much higher than non-peak hour rates.

The originator of the maximum demand system was Arthur Wright of the Brighton supply undertaking. Wright produced a tariff and invented a maximum demand meter. This system was widely diffused and extensively used in the United States and Germany.[4] The supply industries spent the 1890s rearranging their tariffs to give the cheapest rates to consumers with the best load factors. The most important element in achieving a favorable load factor was to induce the user to take current during period of low demand. One of the best ways of doing this was to expand the market for electricity to encompass demand at off-peak generating times. Favorable load factors were the result of diverse loads and the acquisition of a power load was the most important part of a diverse load. Power users needed electricity during the work day and demand for power dropped just at the time when the evening peak lighting and traction loads began. The profitable operation of a central station in the 1890s was dependent on the building up of a power load. German and American supply organizations were quick to recognize this fact.

Although the maximum demand system had originated in England, the English supply industry was a laggard in sales and marketing techniques. It was generally agreed in the Institution of Electrical Engineers that marketing had been ignored in the English Industry.[5] Arthur Wright and Charles Merz were virtually alone in their efforts to court a power load and maintain a close relationship with the customer. It was well known in the English supply industry that the municipal undertakings preferred to let the customer come to them, rather than vice versa. The Merz and McLellan consulting firm concluded that marketing was very poor in the supply industry and "failure to recognise these essentials accounts for the backward state of the industry as a rule, in this country."[6]

The central stations built in the second and third phases were intended to supply light and traction needs, and the English supply organizations had neither the capital nor the inclination to take up a large power load. During the 1890s the major problem facing them was to ensure that they had the capacity to meet the winter peak load. The years 1898 and 1899 were marked by a great many supply interruptions and breakdowns. Partly to blame were delays in delivering the equipment for the many extensions planned to increase the capacity of generating plant.[7] The English supply undertakings had greatly underestimated the demand for electricity and subsequently all additions to capacity were made to meet increases in demand for light. In contrast, the forced development of the German and American supply industries led them to actively court power loads to sustain the great increase in supply capacity.

The unrestrained expansion of the German and American supply organizations during the third phase was partly caused by the necessity of purchasing the products of the manufacturers who were financing the growth of supply systems. The marketing and sales organizations of the electrical manufacturers provided excellent examples for the supply undertakings to follow. The creation of marketing departments in electrical manufacturing came as the result of the intense competition in the industry and from the close ties that existed between manufacturers and their customers. As Chandler noted, entirely new sales organizations were constructed which brought together the customer and a trained engineer who knew much more about the product than the buyer.[8] While English manufacturers relied on agents, German and American firms built up national sales organizations. Their highly trained salesmen had close ties with engineering departments which gave these companies the competitive edge. Excellence in marketing was the key to survival in highly competitive electrical markets and a premium was placed on salesmanship. When the Edison companies were merged with Thomson-Houston to form General Electric in 1892 it was expected that the new company would be organized around the larger Edison company. Yet the bankers, led by J.P. Morgan, were impressed by the superior management and sales department of Thomson-Houston and therefore made it the basis of the new company. The president of General Electric was formerly a shoe salesman, and his expertise in marketing was a factor in his appointment.[9]

The German and American supply companies created sales departments which informed customers about new products and services, and also advised them on engineering matters. A concentrated effort was made to canvass prospective customers and plan future demand schedules. The only English supply company to follow this lead was Nesco. Charles Merz had visited many foreign undertakings and was convinced that "a good pushing job" was essential in central station operations. Nesco canvassers prepared a detailed analysis of the market and each customer was approached in order to assess his electrical needs and tailor his demand into the overall demand schedule. Merz stressed the need for close liaison with the customer, and Nesco was one of the first companies in the English supply industry to open a consumers department.[10]

The specialized market services developed by large electrical manufacturers were instrumental in the diffusion of electric power in industry. The adoption of electric power involved more than the replacement of an existing power source with an electric motor. The industrial user first had to be educated in the use of electric power and his plant had to be radically reorganized to make the most out of a new power source. The marketing of electric power involved not only selling the concept of electric drive, but also advising the customer on his power needs and the best way to employ the electric power he bought. Detailed surveys had to be carried out to determine the needs of the customer and the specific areas where electric power could be most profitably used. The supplier of electrical power had to work hand in hand with the customer to develop the format of power to fit production methods.

As soon as the advantages of electric power were realized, the need arose for engineers well versed in mechanical, production, and electrical engineering. Somebody had to decide what size and type of motor to use, where it was to be employed, and how the speeds and torques of different motors could be coordinated into manufacturing processes. In this way a new branch of engineering, called application engineering, was developed out of the sales departments of the large manufacturing companies who were pioneers in applying electric power to industrial uses. The marketing of electric power required a large, highly trained staff that only the largest manufacturers possessed. The German and American multinationals were able to apply their sales and marketing expertise to the problems of marketing electric power. The large supply organizations were likewise in a good position to develop power marketing departments, and the BEW and Nesco quickly created offices to advise customers and plan their power installations. These offices were similar to the technical offices set up by manufacturers to provide skilled assistance to major customers.[11]

These specialized services were beyond the financial and personnel resources of the majority of English supply and manufacturing firms in the third phase. Furthermore, motors had to be connected to separate circuits and needed different switching which meant that extensive changes had to be made in the cable network. English supply undertakings were reluctant to make these expensive changes.[12] The manufacturers were unable to develop an industrial motor which could withstand the rigors of factory use, and this gave electric drive a bad reputation in English industry. The average English industrial user had become accustomed to overloading his steam equipment and not bothering to keep it clean. While a steam engine will work satisfactorily in a dirty working environment, the early electric motors did not. Overloading an electric motor was normally enough to bring it to a halt and possibly start a fire.

One of the developments of the third phase was the great improvement of the reliability of the electric motor and a reduction in its price. Techniques of mass production were used to produce thousands of standardized motors, and economies of scale meant that one type of General Electric motor that cost \$2,600 in 1891 sold for \$750 in 1895.[13] The lowering of the cost of electric power made the electric motor more attractive to the industrial user at the end of the third phase, and from this time onwards English supply companies began to make an effort to market electric power. A few pioneering undertakings led the way in the marketing techniques of lending out motors, arranging a preferential tariff for power loads, and advising customers on power usage. Bradford was again a pioneering municipal organization, and by 1901 the Bradford undertaking supplied more electric motors than the larger city of Liverpool and slightly less than Manchester.[14] Coventry followed Bradford's lead and its growing skill in marketing electric power played a substantial role in that city's cycle, automobile, and armaments industries.

The success of these policies is evident in the increase in power loads. Once a power load was secured, it became the fastest growing component of demand. The increase of supply organizations' power load was a factor in the quickening pace of factory and workshop electrification after 1900. The bulk supply companies that were founded at this time were based on the plan of achieving a base load in cities and then branching out into outlying areas. The municipalities' opposition to bulk supply in their distribution areas forced the bulk suppliers to develop markets in suburban and rural areas and also pressed them into a program of industrial electrification. The remarkable success of the Nesco bulk supply system proved that the industrial user could provide the base load for a highly profitable enterprise. The Nesco scheme involved the electrification of the mining, shipbuilding, and iron and steel industries in its distribution area. It was rightly called "the Mecca of electrical development" and it provided the example for every bulk system to follow.[15]

The slow adoption of electric power in English industry formed a part of the lag in electrical technology; the fourth phase was the time that most public attention was directed towards industrial use of electricity. Much of the lag was a lag in application engineering because the smaller English electrical manufacturers did not have the resources to provide the specialized services that were vital to industrial power use. During the fourth phase, large bulk supply schemes like Nesco carried out the required application engineering. The great success of Nesco was the result of its low cost electricity and its flair and persistence in marketing. Although application engineering in England was backward compared with its development in Germany and the United States, the fourth phase witnessed some significant progress in this area.

The key to good application engineering was liaison between the electrical engineering of the manufacturer and the mechanical and hydraulic engineering of the customer. A representative of the German electrical industry, speaking at a meeting of the Iron and Steel Institute, commented that German electrical manufacturers "worked hand in hand with tool makers and machine builders ... studying conscientiously the best method in every single case."[16] The records of Nesco show that successful industrial electrification was a long process of evaluation and improvement carried out between the supplier and user of the equipment. Motors, drive systems, and controlling devices had to be adapted to each specific use and the application engineer had to understand the industrial process being electrified. The technology of application engineering was slow to be developed but quick to be diffused. The English bulk supply companies and the electrical manufacturers were adept at copying the products and services of their American and German colleagues. In the fourth phase, American production methods, marketing techniques, and advertising procedures were imitated by English manufacturers. A reverse of the "brain drain" occurred as many American engineers were persuaded to join English firms.[17]

The initial marketing of electric power was carried out in the engineering and shipbuilding industries. The average engineering workshop depended on the precision working of machine tools that were widely spread about the works. In this framework of production the flexibility and economy of electric drive could hardly be ignored. The electrical manufacturers led the way in this field because they had accumulated valuable expertise in the electrification of their own workshops. Their need for high precision and automation gave an impetus to the development of more precise controlling devices and better electric motors, which could also be applied to other engineering works.[18] The first use of electric driving was in the powering of machine tools, particularly turning, milling, grinding, and drilling. As application engineering progressed, more uses were found for electricity, notably through the use of spur gearing which produced vertical motion. The electrification of machine tools started with the use of large electric motors which drove long lines of shafting and ended with the attachment of a small electric motor to each machine.

The figures of industrial power use show that the engineering and shipbuilding industries were the first to adopt electric power on a large scale and were the largest users of electricity in industry. The Factory Inspectors reported in 1902 that electric power was in general use in shipyards and that it provided the great majority of power needs. Yet Sidney Pollard and Paul Robertson have made the point that cyclical demand for ships made entrepreneurs in shipbuilding conservative in their attitudes towards adopting highly capital intensive technology.[19] The experience of the Merz companies on Tyneside indicates that as soon as electrical equipment was perfected for shipyard use, it was enthusiastically adopted.

While equipment for engineering and shipyard electrification was quickly developed, application engineering in coal mining was carried out at a much slower pace, despite the efforts of several electrical manufacturers who concentrated on the electrification of mining. In theory, electrical energy is ideally suited to mining. It is flexible, economical, and can be distributed long distances above and below the surface. In practice, the electrical plant employed in English mines in the second and third phases was ill-fitted to the demands placed upon it. The poor quality of the plant installed in lighting schemes during the second phase was equally evident in mining electrification. Power and lighting equipment in mines was subjected to harsh treatment in wet and dirty conditions and often broke down. The institutional framework of mining in England made breakdowns in hauling and ventilation subject to stringent conditions. For example, the breakdown of ventilation for a short period of time required the mine manager to evacuate all underground workings.[20] Shoddy workmanship and frequent breakdowns did great damage to the reputation of electricity in mining and was a considerable barrier to the electrification of English mines.[21]

There were many problems to be solved by application engineers before electrical equipment could be made suitable to mining. Cables had to be made strong enough to withstand frequent immersion in water, rough handling, and crushing under falls of coal. Electric motors had to be built to handle excessive overloads, resist dust and water, and not spark or catch fire. The early direct current motors used in mining suffered from all these serious disadvantages and helped to give electricity a very bad reputation for safety. English firms that were attempting to bring electric power to mines found out that a strong prejudice existed against electric equipment on the grounds that it was unsafe for underground use. The introduction of the three-phase induction motor solved many problems of application engineering in mining. It was a tough and reliable motor that did not spark and had the added advantage of slowing down to a halt when overloaded which was preferable to it catching fire.[22]

Unfortunately three-phase technology was diffused much slower in English mines than in German and American mines. Several surveys carried out between 1900 and 1905 by mining engineers and the <u>Electrical Review</u> discovered that three-phase power was the predominant system in use in foreign mines. Electricity was generated in large central stations near collieries and steel works and then distributed over wide areas to many mines. The availability of cheap coal and waste heat reduced the price of electric power and induced mine owners to electrify all their operations and become part of a unified distribution scheme.[23] On the other hand, English collieries stayed with direct current equipment and were rarely given the opportunity to join large three-phase schemes. The large power station supplying many mines and ancillary works was rare in English industry, and it was mainly in the integrated concerns of the large engineering and steel manufacturers, like Blockow

Vaughan, that the wide scale distribution system was employed.[24]

The three-phase system was readily incorporated into German and American practice because mining in both countries was carried out in consolidated units. Close geographical location and the ambitions of several large companies aided in the formation of large integrated concerns. In the German Ruhr area in 1893, the ten largest coal mining concerns accounted for one third of total coal output, and by 1910 the ten largest concerns accounted for sixty per cent of output.[25] It was much easier to build a three-phase industrial power system in these integrated units than in the mass of individual companies that composed the English mining industry. In 1913 there were 1,589 separate concerns operating 3,289 collieries.[26] The large bulk suppliers, such as Nesco, became the agents of large scale distribution systems in mining. The Nesco operation was the first to offer three-phase supply to mining. Yet it was not until 1905 that its distribution networks reached out into the Durham coalfields. It took several more years before bulk suppliers in South Wales and the Midlands were able to offer a cheap supply of three-phase current to coal mines.

The records of the Merz companies indicate the problems faced in the marketing of electric power in mines. Many mine-owners were wary of the safety risk attending electric power and refused to consider it. Others did not want to relinquish control over their power source by buying electricity from a supply company, mainly because the breakdown of power supply was extremely costly and electric plant in mines had a poor record for reliability. One of the major objections to electric power was its greater first cost, and this forms a theme in surviving records of engineers who sold electric equipment to mines. The Durham mine owners frequently made the point to Merz's canvassers that even bulk power at its cheapest, between 1d and 1/2d, could not compete with steam costs.[27] The use of direct current equipment must have led to even higher costs for electric power because the user also had to provide his generating plant. Although Nesco converted many Durham mine owners to three-phase use, the major system in use in English collieries was direct current.

The use of electric power in winding is an example of a serious lag in electric power use in England. This was an aspect of mining electrification that was led by the Germans and Americans, with English mines hardly using electric power for the purpose of lifting men and material out of the mine.[28] The application of electric power to mining was a difficult task because the motor had to carry heavy overloads and be capable of rapid build up of torque. The Colliery Guardian investigated this lag and found that the thirty per cent savings in running costs of electrical plant could not compensate for the initial high cost. It was also discovered that the thirty per cent figure claimed by the manufacturers was highly optimistic and that the actual savings were less than the running costs alone. The Colliery Guardian concluded that electric winding would not be employed in English mines in any numbers until the equipment and its control gear was improved and its cost greatly reduced.[29]

The technology used in mines had an interrelatedness that made piecemeal changes uneconomical. For example, the mechanization of cutting at the pit head increased the load on haulage gear and therefore it too had to be mechanized. The more coal that was cut and hauled, the more ventilation and pumping had to be increased. There was an interrelatedness in electric technology used in mines, for one of its major advantages was in the many different uses of a flexible power source. The use of large three-phase systems in Germany and the United States meant that the mine owner was offered a package of all-electric power which could completely replace existing power sources. In contrast, the English mine owner was rarely given this opportunity. The practice in English mining was to electrify in a selective manner. First pumping and ventilation would be electrified (because this was the cheapest to install), and perhaps electric coal cutters or haulage devices would be added some time later.[30] One survey of mine electrification found that electrical equipment was introduced timidly and only temporary installations were made until the equipment proved itself. This led to cheap and slapdash installations which naturally had a high breakdown rate, which in turn reinforced the mine owners' fears that electric plant was unsafe and unreliable.[31]

The system of steam power used in English mines had its own technological interrelatedness which was a barrier to electrification. Cheap "small coals" were used at the pit head in ancient low pressure steam engines. Coal mine managers had a tradition of using antique equipment which was well justified in terms of cost. "Small coals" had such a low value that it was hardly worth the trouble selling it. At less than three shillings a ton this was a very cheap power source. The large amounts of cheap coal available for steam raising was a consequence of the existing methods of haulage and screening, which resulted in large amounts of coal that were not worth selling. An all-electric haulage system would have made steam power at the pit head much more expensive by reducing the amount of "small coals."[32] There were safety and reliability factors to consider in English mines. An old, low pressure steam engine was reliable and would run for long periods without breaking down. Reliability was a paramount consideration in mining for breakdowns were expensive and often dangerous. Coal mine owners preferred using low pressure, non-condensing plant which put less strain on working parts and a long and trouble-free working life.[33] The English mine owner has been given a reputation of technological conservatism which is probably well deserved. Yet until electric plant in mines was improved and reduced in cost, even the best application engineers and salesman could not sell it.

The application engineers in the iron and steel industry faced many of the problems that confronted their colleagues in mining. Motors were overloaded and subjected to habitual mistreatment. Moreover, loads were higher in steel making. The gradual improvement of the electric motor's power and efficiency determined the pace at which electric power was applied to iron and steel. The greater experience of German and American engineers in traction technology gave them the lead in electrification in heavy industry. Their large manufacturing concerns led the way in adapting motors, drive systems, and controlling devices. On the other hand, English manufacturers and application engineers were mainly concerned with light engineering and mining as fields for electric power. One of the most prominent application engineers in England, D. Selby-Bigge, admitted in 1894 that most attention was being given to engineering and mining electrification and that there was no large installation of electric power anywhere in the British iron and steel industry. In 1902 Selby-Bigge was still complaining of a considerable lag.[34]

Electric power was first introduced into iron and steel works in the form of haulage devices and travelling cranes. The latter provided good business for German electrical manufacturers, who found a wide market in travelling cranes in all sectors of industry.[35] From this start a series of electrically driven conveying and charging machines were developed for the larger blast furnaces that were introduced into the German and American mills. The greatest problem of application engineering was found in rolling mills where motors had to carry great overloads in one direction, stop, then quickly reverse. This problem received the attention of several Continental manufacturers and many application engineers until finally a solution was found in the use of balancing converters. The perfection of rolling mill electrification was a testimony to the maturation of application engineering because it led to complex operations being controlled by one man working one lever. There were great economies to be gained in manpower and even the electric drive for rolling mills used less power than steam.[36]

As the pioneering work of application had been carried out in large integrated concerns, the electric power equipment was suited to these conditions. The costs of electrifying steel works were very great because technological interrelatedness of plant required a complete remodelling of plant to suit electric charging, haulage, and rolling. The application of electric drive to steel making was engineered to suit the integrated industrial unit and to exploit the economies of scale. The integration of blast furnaces and coking ovens into three-phase systems was the key to the cheap power that was so readily adopted in the German and American industries. The exploitation of waste gases and heat in steam-raising engines lowered the costs of generation and the use of high voltage technology reduced distribution costs. The technology of waste heat utilization was neglected in England. Much of the research in this important field was carried out on the Continent, and the John Cockerill works at Seraing, Belgium, pioneered the basic process.[37] Waste heat had enormous potential in cheap power supply, and some of the very large steel works of the U.S. Steel company were powered entirely by electricity derived from waste heat and gas engines.[38]

The activities of technological leaders like U.S. Steel received a great deal of attention in the English electrical and steel making press. The new works built by U.S. Steel were all electric and this impressed English observers. Yet it was a whimsical notion to believe that the plants of U.S. Steel could be imitated in the smaller and less integrated English industry. The markets, structure, and expectations of the English steel industry were much different from those in Germany and the United States. The buoyant conditions of the American steel industry were not evident in England, nor was the motivation to expend lavish amounts of capital on labor-saving electrical machinery. In the words of the president of the Iron and Steel Institute, "We have been repeatedly advised to innovate, but to what avail is labor saving machinery, when we already have thousands of willing hands out of work?"[39]

Those elements of the English steel-making industry that

were similar to the large integrated German and American concerns were quick to electrify and followed much the same pattern of adoption that had been begun in the United States. The Dorman Long Company was an integrated steel and heavy engineering concern and it was a pioneer in factory and steel works electrification.[40] Similarly, Armstrong's and Vickers adopted electric drive in their diversified works and mines. Nesco was also active in bringing three-phase power based on waste heat to the iron and steel works of Cleveland and Durham. The advanced technology employed in these concerns has often escaped the attention of scholars who have stressed the backwardness of the British iron and steel industry.

It was acknowledged within the electrical engineering industry that application engineering and the marketing of electric power was the most difficult in the textile industry. Electricity's advantages of flexibility and ease of control were less desirable in the textile industry. The cotton and other textile mills of England used a great amount of power, but power that could be easily transmitted and controlled was not necessary. Textile machinery was run at regular speed throughout the working day, and the most suitable power system for this pattern of usage was large steam engines running long shafts and belts. Electricity could not compete with steam in either suitability or economy of driving. One engineer noted that there would be hardly any business in textiles until either application engineers developed a different kind of electric power system or textile machinery builders introduced new forms of machines.[41] Unfortunately, the former concentrated their efforts in engineering and mining, while the latter designed machines that were suitable for electric driving but sold them overseas.

English application engineers and electrical manufacturers marketed electric power on its merits of increased flexibility and lower costs, and normally held up the American example as proof of the viability of the system. This approach worked in the engineering industry but failed to convince many textile mill owners. The mills of Lancashire and Yorkshire had an excellent reputation for maintaining their power equipment in good order and for getting every possible horse-power from their steam engines. Prospective customers were most concerned with costs and no amount of persuasion could alter the fact that, on average, electric power was more expensive than steam power. The foreign mill owner often had the advantages of water power which could be transformed into cheap electric power. English textile mills were powered by their dependable steam engines and Lancashire boilers, and as Clapham put it, "British factories had their steam engines, belts and shafting in good running order... Coal was cheap."[42]

Isolated plant producing electric power could not compete with steam power in English industry in terms of cost. In some cases, especially engineering, shipbuilding, and steel making, cheap three-phase electricity was provided by bulk supply companies at lower costs than existing isolated or steam plant. In textiles, even the cheapest bulk-supplied current could not compete with steam power. As a general guide line, the bulk supply companies estimated that when current dropped to 1d a unit there would be a wide market for industrial power. The experiences of Nesco, and the Yorkshire and Lancashire bulk systems, indicate that current had to be near to 1/2d per unit before textile mills adopted it. Furthermore, textile users were not good customers for bulk supply concerns because of poor load factors. It was customary in the Yorkshire and Lancashire mills to work steadily through the day and stop at five o'clock. This was because the predominantly female work force was required by law to work these hours. The short working day and the lack of overtime made textile mills unsatisfactory consumers of electric power.[43]

The marketing of electric power in the English textile industry made little headway in the fourth period of electrical progress. Byatt estimates that in 1907 only 5 per cent of the textile industry's power needs were met by electricity.[44] This compares with the figure of 74 per cent in the engineering trades and indicates the uneven diffusion of electrical power in English industry. Poor application engineering hindered the diffusion of electric power in textile mills in the fourth period in much the same way as it did in iron and steel and mining in the third.

Where power was applicable and well marketed, it was soon adopted. Where application engineering was poor and alternative sources of power were cheaper, it was not.

CHAPTER NINE

THE LAG IN ELECTRICITY RECONSIDERED

The lag in electrical technology in England was mainly in the minds of the Victorians whose perception of the lag was influenced by many different factors. Perceptions of the lag were not based on an appreciation of the pace and development of electrical technology, but instead reflected a wide range of special interests and prejudices. A lag was perceived because electrical development was measured against a carefully chosen yardstick of foreign installations. If the methods of measurement had been different, and England compared to other nations, such as France or Belgium, the lag would have disappeared. The lag served many interests, and many masters, and it was therefore useful in the political environment of technological change. If perceptions of the lag had not existed by the first phase of electrical progress, they would have been created.

Perceptions of the lag in electrical technology were based on a selective interpretation of the English situation and an over-estimation of the advances of foreign systems which were used in comparisons. The English supply system was alleged to be a "patchwork" of small, inefficient undertakings which hindered the formation of a national grid. The German supply system was also a patchwork, with many small and inefficient undertakings. For example, there were fifty different voltages for power use and nearly sixty different voltages for electric lighting.[1] Many of the comparisons made between the English electrical industry and those in Germany and the United States pitted the worst of the English with the best of the foreign systems. As the electricity supply of London was a heated issue during the fourth phase, comparisons of London with Berlin or New York were commonplace. Figures produced by the bulk supply companies showed that per capita horsepower of electricity connected for power and light was 26 in London compared with 73 for New York and 82 for Berlin. The chaotic supply of London made it easy to show a lag, but if the supply of Newcastle had been chosen to represent the English supply industry the lag would not have been evident. Newcastle had 115 horsepower of electricity connected for light and power per capita, which made it one of the world's leading electrical undertakings.[2]

It was the perceptions of the lag, and not the real situation in the electrical industry, that were discussed in England. This study has shown that the reality of electrical development was rarely captured in the depiction of the lag.

Contemporaries placed the beginning of the lag in the second phase, yet it had really begun in the first. Despite beliefs that the second phase marked the decline of electrical progress in England, this period saw much of the lost ground recovered by the English industry. Perceptions of the lag reached a climax during the third phase when attention was concentrated on two highly visible technologies, multi-phase alternating current and electric traction.

The invasion of large foreign companies in the 1890s added fuel to the fire, for these companies were strong in traction and multi-phase technologies. The demise of the English electrical manufacturers was deemed to be imminent. When two of the largest electrical manufacturers (Westinghouse and General Electric) announced that they were going to build factories in England at this time, the response of the English electrical industry was pessimistic. The new factories were to incorporate the newest manufacturing equipment and production techniques, and spokesmen of the English industry feared that a small number of large, well-organized American companies would dominate the English market. They commented that the English electrical manufacturers "stood aside and waited with baited breath."[3]

Perceptions of the lag dominated the discussions of legislators and electrical engineers. The press reported the American "invasion" in gloomy editorials. Yet the fears of foreign domination of the English industry proved to be groundless. Westinghouse and GE did build very advanced factories and did corner the market in some specific areas, especially traction and large generating units for central stations. Their technological superiority was in the heavy engineering of the third phase of electrical development. This advantage waned in the next phase because the great traction and supply schemes were quickly completed and the demand for heavy electrical engineering decreased.

Westinghouse and GE did not make the great profits that were expected, and initially they faced severe financial problems because of the unprofitability of their English operations. The great showpieces of electrical development, like the London Underground Railway, increased perceptions of the lag but did not produce a return on the massive amounts of capital invested in them.[4] In the meantime, the advanced techniques of foreign manufacturers were copied by English firms and the new American-designed factories became the models of a host of imitations. By the fourth phase English electrical manufacturers were using mass production techniques in their "American shops," keeping standardized equipment and parts in stock, and improving their sales and delivery performance.[5] The Westinghouse factory at Trafford Park, near Manchester, was the pilot installation of a modern industrial "park" which Hudson has called "the cradle of the Second Industrial Revolution" in England.[6]

The English electrical industry was able to learn and profit from the foreign "invasion" of the third phase and this contributed to their strong showing in the fourth phase. The electrical market had shifted from heavy engineering to consumer durables. The consumer appliances and small electrical components of the fourth phase favored the smaller electrical companies who mass produced electrical goods for specific markets. New customers were found for electric heating and cooking, and the new automobile and aviation industries needed precision machine tools powered by small electric motors in addition to millions of electrical components for their products.[7]

These markets were easy to enter, high in volume, and provided excellent sales for the smaller electrical manufacturers. Great efforts were made by English companies to improve their production techniques and marketing organization. American practices were highly regarded and universally copied. The supply companies of the United States emerged from the era of consolidation with a greater emphasis on better management and marketing. The meetings of the National Electric Light Association during the fourth phase were dominated by discussions of the ways to increase consumption of electricity. The development of a power load and the marketing of electrical appliances were given a high priority and it was agreed in these meetings that marketing depended on an "aggressive and persistent effort" in promotion, advertising, and sales. The supply companies increased their liaison with both manufacturer and customers, built up their advertising programs, and began demonstrating and installing appliances. From about 1904 onwards, the American supply industry began "systematic efforts to bring our wares to the attention of the public." The opening of "commercial" departments brought a new emphasis on acquiring and training the sales personnel that were essential to mass marketing.[8]

The new consumer-oriented technology of the fourth phase put a premium on commercial expertise and application engineering. Much of the discussion of the lag in the preceding phase was centered on the merits of advanced scientific education which had supposedly given foreign engineers an advantage in a new "scientific" industry. This debate continued in the fourth phase even though the personnel requirements of the electrical industry had changed. The press and general public still believed that advanced scientific education was the foundation for success in the electrical industry. The focus of educators and electrical engineers was different. The conclusions drawn at the Second Annual Conference on Education and Training of Engineers (1911) emphasized the advantages of a practical education which encompassed many areas of engineering and also included business and language courses. The German electrical industry was still held up as the model to follow, but it was the German engineers' commercial acumen and skill with languages that was admired, and not their scientific training.[9]

The Conference called for a more varied and broad-based education for engineers that would include practical experience and theoretical elements. The need for engineers well-versed in several branches of engineering and possessing commercial and business skills was emphasized by delegates at the Conference and by successive presidents of the I.E.E.[10] The electrical industry needed salesman, business managers, and application engineers in the fourth phase. Professional electrical engineers knew that the college-trained man was less likely to get a good post in England than in the United States and Germany. This was also the conclusion of the 1908/9 Report of the Board of Education, which blamed the slow growth of technical education on "the small demand... for young men trained in the theoretical side of industrial operations and in the sciences underlying them."[11] In view of this situation, a more practical education was welcomed by electrical engineers in England.

The importance of marketing was recognized by engineers and entrepreneurs in the English electrical industry in the fourth phase. The speeches and writings of the spokesmen of the industry stressed the need to copy German and American "push" and marketing strategies. The highly trained workmen in the industry, the skilled craftsmen who allegedly gave the German companies the competitive edge, were not so highly prized in the twentieth century. Many manufacturing processes were standardized and adapted for non-skilled labor. The German and American electrical manufacturers had led in this development and they were followed by the English after 1900. In the twentieth century it was the "commercial men" and skilled managers who were most needed. The electrical engineers who spoke at the I.E.E., and the company directors who wrote in the electrical press, praised the English workman as good as the best in Germany and the United States; it was English salesmen, application engineers, and managers who were criticized as being inferior. The essence of electrical progress after 1900 was in expanding the market to encompass a new range of electrical goods and services made possible by a much cheaper supply of electricity. As one president of an I.E.E. subsection put it, "the lag is commercial rather than technical."[12]

Despite the efforts of a small number of undertakings, the third phase marked English companies lagging behind German and American firms in marketing techniques. The fourth phase saw a rapid development of the technologies of mass marketing which English companies were much quicker to adopt. The consumer revolution of the fourth phase was based upon the exploitation of a new market for electrical goods and services. In the supply industry, efforts were made to bring electric light and heat to areas and customers that had not been approached in the previous phases. In England the great expansion of the supply system, which began around the turn of the century, brought the distribution networks into areas that were working class rather than residential. As the distribution networks followed the tramway lines into manufacturing districts, the number of potential customers increased. The "super" power stations were built to meet increasing traction loads and had excess capacity that could be used to supply electric power users. An _Electrical Review_ survey of 1900 reported that many supply undertakings were more actively marketing power equipment and services, and also were courting a power load. The survey concluded that interest in electric power was on the increase.[13]

The English supply industry was quick to appreciate the consumer orientation of the fourth phase of electrical development. Like their American counterparts, they had built very large stations and needed very large levels of demand to operate profitably. The traction and lighting markets were saturated by the end of the third phase, consequently the next phase saw attempts to develop new markets and new products. The English supply industry kept a close watch on the marketing techniques being developed in the United States. By 1906 English supply organizations were creating "commercial" departments and developing systems of mass marketing.[14]

The new marketing techniques employed in the English supply industry were copied from the American example and from the gas industry. Programs of advertising were initiated, houses were wired without cost, motors and appliances were loaned out or sold in installments, and a staff of salesmen, business managers, and canvassers assembled.[15] The most visible sign in the new marketing techniques was the supply company "showroom" which advertised electrical goods and services, demonstrated appliances, and arranged for free inspection and installation in consumers's premises. These showrooms were established on the "high street" and operated like a retail store. The municipal lighting operations, which ignored marketing in the third phase, quickly began to adopt the marketing techniques of the fourth phase. They opened showrooms, hired commercial staff and armies of canvassers, and took an active part in advertising and promotion. As the chairman of the Ilford municipal undertaking said, "the council, like other commercial people, must advertise their business."[16]

The English electrical industry made impressive progress during the fourth phase, and made up much of the ground lost in the nineteenth century. The average size of new stations built in this period increased dramatically; the new "super" stations incorporated the latest power and generating technology. The English supply system was efficient and technologically advanced in the fourth phase of electrical progress, yet it was still perceived as lagging behind best practice. In terms of the size of the supply system, indicated by kilowatts connected, the German industry moved ahead of the English during this phase. The German electrical association estimated that new central stations were being built every week during this period. According to English engineers who visited Germany in 1901 the German central stations incorporated the latest advances in electrical technology, especially in multi-phase current in which the English supposedly lagged. The new stations built in Germany during the fourth phase were not the advanced "super" stations; on the contrary, they were small direct current stations that represented a technology that was obsolete by the 1890s.[17] Throughout the fourth phase the average size of German stations grew smaller while the average size of English installations grew larger.

Much of the increased capacity in the German and American industries during the fourth phase came from new services to smaller towns and rural communities.[18] In contrast, these markets got less attention in England because the growth of capacity took the form of extensions to the large urban systems. In terms of the technology of public supply, the progress of the fourth phase was towards larger power stations supplying large areas with light and power. The development of the English supply system from the turn of the century onwards was in the areas of bulk supply and large municipal systems, and consequently large numbers of customers were supplied with electricity at lower prices than Germany or the United States.

Despite criticisms to the contrary, the municipal undertakings in the supply industry performed adequately during the fourth phase. Great progress was made in the construction of electric tramways. Although the English traction system was not as large as that in the United States, 95 per cent of the total tramway system was electrified by 1914.[19] The municipalities added to their power stations and built new ones. Their low prices for electricity brought in more customers. Municipal supply and traction undertakings prospered in the fourth phase. Their successful acquisition of a power load, and their policy of annexing smaller municipal neighbors, made them larger and more efficient generating units.

The municipalities made every effort to strengthen their financial and political position by making every effort to market cheap electric power because the institutional framework of bulk supply would not allow private companies to operate in municipal areas until the latter were shown to be unable to supply cheap power. As long as large municipal undertakings in cities like Manchester, Sheffield, Liverpool, and Birmingham were able to supply low cost electricity to industry, their position was secure from the bulk power systems which constituted the supply technology of the future.[20]

The great progress made in public supply in England during the fourth phase went unnoticed as most attention was on the issues of bulk supply and industrial power. After 1900 the nature of the lag changed because a new phase of electrical development had begun. New electrical technologies became the areas in which contemporaries believed England was lagging. Unfortunately, the great progress of the British electrical industry during the first decade of the new century was obscured by public discussion of lags in newer technologies. If the municipal undertakings had remained as small and inefficient as their critics liked to make out, they would have succumbed quickly to the competition of the bulk suppliers. In fact they remained competitive during the fourth phase and thus proved difficult to incorporate into larger systems when efforts were made in the 1920s and 1930s. The propaganda of the bulk supply companies continued to depict the supply industry as small scale, unstandardized, and inefficient.

The reports of several government committees, which investigated the supply industry after the First World War, reinforced this viewpoint. The highly influential report of the Weir Commission of 1926 underlined the weakness of England's supply system and blamed it for the poor record of electrification in English industry. It was alleged that the high cost of electricity retarded the growth of electric power usage in England.[21]

Many historians have repeated this argument and substantiated it with claims of expensive electricity in England. Ensor argues that high electricity prices in England retarded industrial development. Although Ensor argued that prices for electricity were much higher in England than on the Continent, his estimated average price of between 6d and 8d per kilowatt hour is much greater than the 2d average price estimated in 1902. The average price per unit in London was around 3d which was extremely high for English undertakings, and a consequence of the many small undertakings operating in the metropolis. On the other hand, the price per unit of electricity in the large municipal systems in the provinces and in the bulk supply areas started the fourth phase at just over 1d per unit and dropped to around 1/2d by 1913.[22] English supply undertakings were often able to deliver current to industry below 1d per unit after 1900.

The Weir Committee's technological viewpoint was influenced by the large-system orientation of its consultants, and its report was used as a platform to argue the advantages of a large integrated system founded on a small number of very large stations.[22] A comparison with the much vaunted German system of electricity supply reveals that English power stations were, on average, growing larger and more efficient during the first fourteen years of the twentieth century. The record of the English supply industry during the First World War was far from unsatisfactory; despite losses of skilled personnel and the greatly increased costs of inputs, the industry managed to double its output during the war.[24] This impressive expansion came as a result of the increased size and productivity that the industry successfully achieved in the fourth phase of electrical progress. It can be interpreted as the forced expansion of a system that was operating at a highly efficient level in 1914.

In 1911 the president of the Manchester section of the I.E.E. noted that "at no time in the history of electrical engineering was there as much activity as there is today."[25] In spite of great technological progress and excellent business, there were still claims of a lag in the industry. A survey of the electrical manufacturers, carried out in 1911, recorded that they were far from happy with their lot. They claimed that foreign competition was damaging the industry and portrayed the situation in the most pessimistic of terms. The solution to these problems were sought in active government intervention, in the form of tariffs, increased taxation of multinationals operating in England, and a government-sponsored "Electrical Bank" which would emulate the financial backing of the German and American industries.[26]

The portrayal of a lag in the English electrical industry was an important part of the manufacturers' campaign to improve their competitive position. As long as the perception of the lag served somebody's interest, it would continue to be discussed and disseminated within the electrical industry. All sections of the industry had something to lose, or gain, by government intervention. The institutional framework of the supply industry made the government, represented by the Board of Trade, the referee between the two sides.[27] A minor change in wiring rules or safety regulations had a great impact on the industry and gave an incentive to participate in the decision making process. The Act of 1882 began the process of lobbying and negotiation which helped form the image of lag in the electrical industry. The process was still going on in 1914.

ENDNOTES FOR CHAPTER ONE

1 Walt W. Rostow, The World Economy: History and Prospect (Austin, Texas: University of Texas Press, 1978), p. 165.

2 Peter Mathias, The First Industrial Nation: An Economic History of Britain, 1700-1914 (New York: Charles Scribner's Sons, 1969), p. 417.

3 J.H. Clapham, An Economic History of Modern Britain, Vol. III: Machines and Economic Rivalries, 1887-1914 (Cambridge: University Press, 1938), p. 156.

4 Thomas Parke Hughes, "British Electrical Industry Lag, 1882-1888," Technology and Culture, III (1962), 38.

5 Leslie Hannah, Electricity before Nationalisation (Baltimore: Johns Hopkins University Press, 1979), p.39.

6 Keith Hutchinson, The Decline and Fall of British Capitalism (New York: Charles Scribner's Sons, 1950), p. 9.

7 Brian Murphy, A History of the British Economy (London: Longman's, 1973), p. 695.

8 Richard Tames, Economy and Society in Nineteenth Century England (London: George Allen and Unwin, 1972), p. 57.

9 Mathias, First Industrial Nation, p. 417.

10 Takuji Sakamoto, "Introduction and Development of Technology in the British Machinery Industry, 1875-1916," paper given at the First International Conference of Business History held at Shizuoka, Japan, January 1979, p. 3.

11 Derek De Solla Price, "A Theoretical Basis for Input-Output Analysis of National R&D Policies," in Devendra Sahal, ed., Research, Development and Technological Innovation (Lexington, Mass.: D.C. Heath, 1980), p. 257.

12 Stephen Davis, The Diffusion of Process Innovations (Cambridge: University Press, 1979), pp. 10-14.

13 William Ashworth, A Short History of the International Economy since 1850 (London: Longman's, 1962), p. 6.

14 Jacob E. Goldman, "Science for Economic Growth and Social Change," in W. Dale Crompton, ed., The Interaction of Science and Technology (Urbana: Universiy of Illinois Press, 1969), p. 9.

15 R.C.K. Ensor, England, 1870-1914 (Oxford: Clarendon Press, 1960), p. 508.

16 M. St.J. Parker and D.J. Reid, The British Revolution (London: Blandford Press, 1972), p. 311.

17 H.H. Balin, The Organization of Electricity Supply in Great Britain (London: Electrical Press, 1946), p. 12.

18 Devendar Sahal, "Technological Progress and Policy," in D. Sahal, ed., Research, Development and Technological Innovation, p. 174.

19 Zvi Griliches, "Hybrid Corn: An Exploration of the Economics of Technological Change," Econometrica, XXV (October, 1957), 501-502. Edwin Mansfield, "Technical Change and the Rate of Imitation," Econometrica, XXIX (October, 1961), 741-766.

20 John P. McKay, Tramways and Trolleys: The Rise of Urban Mass Transit in Europe (Princeton: Princeton University Press, 1976), p. 244.

21 Richard A. Hennessey, The Electrical Revolution (Newcastle-on-Tyne: Oriel Press, 1971), p. 26.

22 D.S.L. Cardwell, Technology, Science and History (London: Heineman, 1972), pp. 190-193.

23 A.L. Levine, Industrial Retardation in England, 1880-1914 (London: Weidenfeld and Nicolson, 1969), p. 145.

24 David Landes, The Unbound Prometheus (Cambridge: University Press, 1969), p. 343.

25 Charles A. Bennett, A History of Manual and Industrial Education, 1870-1917 (Peoria, Ill.: Manual Arts Press, 1937), p. 211.

26 I.C.R. Byatt, The English Electrical Industry, 1875-1914 (Oxford: Clarendon Press, 1979), pp. 187-188.

27 Sir Henry Self and Elizabeth Watson, _Electricity Supply in Great Britain_ (London: George Allen and Unwin, 1952), p. 3.

28 Hugh Aitken, _Syntony and Spark: The Origins of Radio_ (New York: Wiley, 1976), pp. 8-9.

NOTES FOR CHAPTER TWO

1 P.F. Mottelay, Biographical History of Electricity and Magnetism (London: C. Griffin, 1926), pp. 174, 246-48.

2 H.C. Oersted, The Discovery of Electromagnetism made in the Year 1820 (Copenhagan: Larsen Reprint, 1920).

3 L. Pearce Williams, "Michael Faraday," in Charles Coulston Gillespie, ed., Dictionary of Scientific Biography (New York: Charles Scribner's Sons, 1972), p. 533-534.

4 L. Pearce Williams, Michael Faraday: A Biography (New York: Basic Books, 1965), pp. 161-183.

5 Quoted in Harold Sharlin, The Making of the Electrical Age (London: Abelard Schuman, 1963), pp. 139-140.

6 Charles Singer et Al., A History of Technology (New York: Oxford University Press, 1958), III, 180.

7 Percy Dunsheath, A History of Electrical Power Engineering (Cambridge: M.I.T. Press, 1962). p. 105.

8 A Chronological History of Electrical Development (New York: National Electrical Manufacturers Association, 1946), p. 43.

9 Malcolm MacLaren, Rise of the Electrical Industry during the Nineteenth Century (Princeton: Princeton University Press, 1943), p. 114.

10 Engineering, XVIII (1874), 414.

11 Edison papers, (West Orange, N.J.), Laboratory Notebook # 781122, pp. 19-24.

12 Siegfried Von Weiher and Herbert Goetzler, Weg und Werken der Siemenswerk in Fortschritt der Elektrotechnik, (Munich: Brukman, 1972), pp. 35-37.

13 Telegraphic Journal and Electrical Review, VI (June, 1878), 293-95. Henceforth this journal will be cited as T.J.E.R.

14 Chronological History of Electrical Development, pp. 43-44.

15 T.J.E.R., VII (January, 1879), 10.

16 Harold Passer, The Electrical Manufacturers, 1875-1900 (Cambridge: Harvard University Press, 1953), p. 121.

17 Electrical World, IV (May, 1884), 160.

18 David Woodbury, Beloved Scientist: Elihu Thompson (New York: MacGraw Hill, 1944), p. 154.

19 Proceedings of the National Electric Light Association, Second Annual Convention (Baltimore: Baltimore Publishing Co., 1886), p. 453.

20 Passer, Electrical Manufacturers, p. 117. Edison Electric Light Companies, Bulletins, 1882-1886.

21 Nicholas Wainwright, History of the Philadelphia Electric Company (Philadelphia: Philadelphia Electric Co., 1961), p. 32.

22 Passer, Electrical Manufacturers, p. 150.

23 Electrotechnische Zeitschrift, XII (January, 1890), 69.

24 Electrical Review, XXX (February, 1892), 194.

25 W.E. Swale, Forerunners of the Northeastern Electricity Board (Manchester: N.E.B., 1963), p. 26.

26 Statistics from The Electrician, Electrotechnische Zeitschrift, and Electrical World.

27 H.G. Prout, A Life of George Westinghouse (New York: Arno Press, 1972), p. 100.

28 The Electrician, XXIV (March, 1890), 222-223.

29 Escher Wyss, 1805-1955: 150 Years of Development (Zurich: Escher Wyss, 1955), p. 33.

30 Prout, George Westinghouse, p. 72.

31 Electrical Review, LVI (May, 1905), 779-780.

32 Merz and McLellan Papers, vol. 1/1.1, 1898, Preliminary Report on Walker and Wallsend Project.

33 Brian Bowers, "Electric Traction: Early Electric Motors," Institution of Electrical Engineers 2nd Historical Weekend, July 1974, London.

34 Frank Sprague, "Some Personal Notes on Electric Railways," in Electrical Review, New York, XL (February, 1902), 226.

35 Harriet Sprague, Frank J. Sprague and the Edison Myth (New York: William Frederick, 1947), pp. 10-11.

36 Electrotechnische Zeitschrift, XI (August, 1890), 446.

37 Siemens and Halske, Electrische Bahnen (Berlin: Siemens and Halske, 1896), pp. 5-8.

38 Figures from Electrical Trades Directory, Electrical World, and Electrotechnische Zeitschrift.

39 Neil B. Reynolds, "G.E. Affiliates in Great Britain," Research Paper, General Electric Research Unit, Schenectady, New York, 4 March 1966.

40 Alfred D. Chandler, The Visible Hand: The Managerial Revolution in American Business (Cambridge, Mass.: Harvard University Press, 1977), pp. 193-194.

41 Byatt, British Electrical Industry, p. 138.

42 Ibid., p. 29.

43 The City and South London Railway was begun in 1890, which places its development in the third phase of electrical progress; The Electrician, XXVI (November, 1890), 9-12.

44 The North Eastern Railway began electrification in 1902; see John Rowland, Progress in Power (London: Newman Neame, 1966), pp. 30-33.

45 T.C. Barker and Michael Robbins, A History of London Transport, Vol. II: The Twentieth Century (London: Allen and Unwin, 1974), pp. 77-80.

46 Ibid., pp. 58-60.

47 Hannah, Electricity before Nationalisation, pp. 161-169.

48 Great Britain, Parliamentary Papers, cd. 1112, vol. xii (1902) "Annual Reports of the Chief Inspector of Factories and Workshops," 1901, p. 3.

49 <u>Ibid.</u>, cd. 2139, vol. x (1904), pp. 100-101, 250-253.

50 <u>Ibid.</u>, cd. 3036, vol. xv (1906), p. 331.

51 Byatt, <u>British Electrical Industry</u>, p. 80.

NOTES FOR CHAPTER THREE

1 J. Munro, "New Sources of Electric Power," Nineteenth Century, XXXVI (1894), 919-921.

2 This issue has been discussed at great length in S.B. Saul, The Myth of the Great Depression, 1873-1896 (London: MacMillan, 1969).

3 Great Britain, Parliamentary Papers, Vol. XXIII, (1886), x, xxiv, "Final Report of the Royal Commission Appointed to inquire into the Depression of Trade and Industry."

4 The Times (London), January 7, 1899, p. 9.

5 The electrical press reflects an increasing concern with Great Britain's foreign policy and noted the deterioration of relations with Germany. The Boer War was seen to have an enormous effect on all areas of public opinion, and "the recent ordeal of national pride" was expected to "shake up" the country including the electrical industry. Electrical Review, XLVII (September, 1900), 490.

6 G.M. Young, Portrait of an Age (London: Oxford University Press, 1969), p. 164.

7 Herman Ausubel, The Late Victorians (New York: Van Nostrand, 1955), pp. 21-22, 111-113.

8 Ibid., pp.23-24.

9 Cassier's (March, 1901), 374-386, contains a series of vewpoints of engineers and industrialists on "American Competition in the World's Engineering Trades."

10 Young, Portrait of an Age, p. 27.

11 An example of this view is Sir A. Noble, "The Need for Technical Education: A Means to Industrial Supremacy," in Cassier's Magazine, XVII (December, 1899). 153.

12 The Economist, XL (May, 1882), 641.

13 Gertrude Ferranti and Richard Ince, The Life and Letters of Sebastian de Ferranti (London: Williams and Norgate, 1934), p. 108.

14 Some of the expert witnesses brought before the Select Committee on Electric Lighting believed that many component parts of the lighting systems then available were at the point of perfection. See The Electric Lighting Act: Minutes of Evidence (London: Scientific Publishing Co., 1882). The Journal of Gas Lighting, XXXV (July, 1880), 130, made the jaundiced comment that the claims of the promoters of the electric light constituted the "silly season" in the press.

15 T.J.E.R., X (April, 1882), 283.

16 This conclusion has been attributed to William Preece, but has been repeated by several other electrical engineers; see Hennessey, Electrical Revolution, p. 47.

17 Great Britain, Parliament, Hansard's Parliamentary Debates, 3rd series, Vol. 36 (1887), 861.

18 Journal of the Society of Telegraph Engineers and Electricians, XV (May, 1886), 62. Henceforth this journal will be cited as J.S.T.E.E.

19 The Times (London), February 12, 1886, p. 14.

20 Great Britain, Parliament, Hansard's Parliamentary Debates, 3rd series, Vol. 36 (1886), 147.

21 Report of the Marindin Commission. Great Britian, Parliament, Parliamentary Papers (House of Commons and Command) 1889, vol. lxx, Cmnd. 189.

22 One of the most publicized accounts of American development was written by William Preece. "Electric Lighting in America" was published in the electrical press, T.J.E.R., XIV (December, 1884), 470; and formed the basis of a session at the Society of Arts, and was later published in vol. XXXIII of their journal.

23 Great Britain, Parliament, Hansard's Parliamentary Debates, 3rd series, Vol. 36, (1886) 140-47. See also his correspondence to The Times, February 3, 1886, P. 6.

24 Robert Hammond was a leading promoter of the House to House Electricity Supply Companies and had an interest in the Fowler Manufacturing Company. Allegations were made that his reports as an independent consulting engineer was favorable to these companies. See Electrical Review, XXXIII (September, 1893), 241.

25 Great Britain, Electric Lighting Act, 1888. 51 and 52 Vic. 1, ch. 12

26 Pall Mall Gazette, January 12, 1882, p. 3.

27 Journal of the Institution of Electrical Engineers, XIX (May, 1890), 448. Henceforth this journal will be cited as J.I.E.E.

28 Electrical Review, XLVIII (June, 1901), 1061.

29 Electrical Review, XXXV (August, 1894), 226-27.

30 J.I.E.E., XXIX (April, 1901), 662.

31 Young, Portrait of an Age, p. 164.

32 Quoted in D.S.L. Cardwell, The Organisation of Science in England (London: Heineman, 1957), p. 101 Commented upon in The Times (London), December 11, 1879.

33 The Times (London), January 20. 1897, p. 9.

34 J.I.E.E. XXX (February, 1901), 475.

35 Fred A. McKenzie, The American Invaders: Their Plans, Tactics and Progress (1896) quoted in Ausubel, The Late Victorians, pp. 25-27.

36 Electrical Review, XLVIII (May, 1901), 807, describing the views of a group of aldermen from Sheffield, who had just returned from a visit to the United States.

37 J.I.E.E., XXXI (January, 1902), 536.

38 Electrical Review, XLVI (March, 1900), 433-34, 477, notes an exchange between a correspondent who claimed that the average charge in Manchester for light was 5d per unit and 1 1/2d per unit for power compared with the average American charge of 1/2d per unit. This claim brought a quick response from Dr. C.H. Wordingham, the chief engineer of the Manchester undertaking, who pointed out that 5d per unit was the highest possible charge on the sliding scale, and that the average price per unit was 3 1/3d. The 1/2d per unit in the United States was dismissed as erroneous and Wordingham added that such a low price could only be achieved with hydroelectric plant and not from a steam powered central station.

39 Electrical Review, XLVIII (March, 1901), 534. A leading article entitiled "The British Press and British Industry" appeared a few weeks later in the issue of May 10,

1901. It commented that "the newspapers have taken it upon themselves to educate the public up to the point of using foreign made plant" (p. 781).

40 *Ibid*., XLVII (September, 1900), 488.

41 C.S. Vesey Brown, "Electrical Supply in Great Britain," *Cassier's Magazine*, XXI (December, 1901), 162.

42 London County Council Archives, CL/HIG/1/74, Printed Document of the Highways Committee, Reports of the Parliamentary Committee, 1907.

43 M.B. Mountain, "Use of Electrical Energy in Mines and Collieries," *Electrical Review*, LVI (May, 1905), 799.

NOTES FOR CHAPTER FOUR

1 This quotation appears in a letter published in the *Electrical Review*, XLIX (September, 1901), 415.

2 In 1898 the average price per kilowatt-hour unit was 4.35d for the municipal undertakings and 5.57d for private companies; *Electrical Review*, XLVI (June, 1900), 1073, quoting the Presidential Address of the Municipal Electrical Association. The municipal tramway undertakings were also charged with undervaluing depreciation costs or even not including them in yearly accounts. See McKay, *Tramways and Trolleys*, pp. 188-89.

3 London County Council Archives, CL/HIG/1/74, Printed Documents of the Highways Committee, Report of the Parliamentary Committee, 1907.

4 *Journal of Gas Lighting*, XLI (February, 1883), 313.

5 Pamphlet accompanying the *Vio-Ray* on exhibit at the Milne Museum of Electrical History, Tonbridge, Kent.

6 Appleyard, *History of the I.E.E.*, p. 89.

7 Josef Redlich and Francis W. Hirst, *The History of Local Government in England* (London: MacMillan, 1970), p. 103.

8 John H. Garvin, *The Life of Joseph Chamberlain* (London: MacMillan, 1932), I, 213.

9 *Journal of Gas Lighting*, XXXV (February, 1880), 236.

10 Patterson, "Gas Companies of London," *Ibid.*, XLV (June, 1885), 10.

11 William B. Munro, *The Government of American Cities* (New York: MacMillan, 1920), p. 16.

12 Ernest S. Griffith, *The Modern Development of City Government* (London: Oxford University Press, 1927), p. 460.

13 Frank J. Goodnow, *City Government in the United States* (New York: Century, 1904), pp. 36-39.

14 Wallace Rice, *Seventy-Five Years of Gas Service in Chicago* (Chicago: People's Gas Company, 1925), p. 5.

15 Charles Zeublin, *American Municipal Progress* (New York: MacMillan, 1916), p. 66; *Journal of Gas Lighting*, LXVIII (May, 1912), 519.

16 Louis Stotz, *The History of the Gas Industry* (New York: Stettiner, 1938), pp. 20-21.

17 Frederick C. Howe, *European Cities at Work* (New York: Charles Scribner's Sons, 1914), p. 113, 330.

18 William Dawson, *Life and Government in Germany* (London: Longman's and Green, 1914), p. 216.

19 Hans-Joachum Braun, "Old and New Technologies: The Competition between Gas Lighting and Electric Lighting in Germany, 1880-1914," unpublished manuscript, Universitat Bochum, 1979, p. 2.

20 Howe, *European Cities at Work*, p. 113.

21 *Electrical World*, XXXVII (January , 1901), 161.

22 Garvin, *Joseph Chamberlain*, I, 188.

23 Derek Fraser, *Power and Authority in the Victorian City* (London: St. Martin's Press, 1979), p. 172.

24 Great Britain, Parliament, *Hansard's Parliamentary Debates*, 3rd series, vol. 36 (1886), 861.

25 *The Law Relating to Electric Lighting*, p. 99.

26 *Ibid*., pp. 100-105.

27 *Electrical Review*, XXX (June, 1892), 26. Under provisions of the Act a local authority could delay a company's application for a provisional order by getting one of its own and not carrying it out. By 1892 there were about 100 municipalities with provisional orders which had not been used to commence supply.

28 Ferranti Archives, Box L7, Proceedings of the London Electricity Supply Company, March 31, 1890, p. 44.

29 *The Electrician*, XXVI (February, 1891), 275.

30 Bristol City Archives, 17845/1; D.G. Tucker, "W.H. Preece, Nineteenth Century Telegraph, Telephone and Power Station Engineer," I.E.E. Historical Paper, 1974, p. 10/7.

31 Ferranti Archives, B 14, The Ferranti System of Electric Lighting (London: Whitehead, Morris and Lowe, 1882). R. Silverberg, _Light for the World: Edison and the Power Industry_ (Princeton, N.J.: Van Nostrand, 1967), p. 186.

32 _Electrical Review_, XXXIII (July, 1893), 86.

33 _The Law Relating to Electricity_, pp. 65-68.

34 Archives of the I.E.E., Folder code Naest 11/14, Papers of the Electrical Trades section of the London Chamber of Commerce. One section of the Electric Lighting Act of 1888 brought all wires outside the jurisdiction of the 1882 Act under Board of Trade control. The companies that had strung overhead wires before 1888 were given one year to remove them, _Electrical Trades Directory for 1893_, p. 87.

35 _T.J.E.R._, XVI (January, 1885), 23. _J.I.E.E._, XXV (January, 1896), 120. It still costs three to ten times as much to lay a cable underground, letter to the author from Robert L. Jones, Information Dept. of Columbus Electricity Company, June 13, 1980.

36 Wainwright, _History of the Philadelphia Company_, p. 15.

37 Kenneth Fox, _Better City Government: Innovation in American Urban Politics_ (Philadelphia: Temple University Press, 1977), p. 22.

38 _T.J.E.R._, XIV (December, 1884), 472.

39 _The Electrician_, XXIV (December 1899). 120.

40 _Electrical World_, II (November, 1883), 190. There was no lack of attempts to regulate or eliminate overhead wires, but the companies were always able to avoid this by applying to a higher authority and evading existing regulations. Zeublin, _American Municipal Progress_, pp. 62-63.

41 Society of Telegraph Engineers and Electricians, _First Wiring Rules_, June 21, 1882. These rules were motivated by fears that unsafe installation would give electricity a bad name. The institution's rules have expanded with the development of electrical technology and are now in the fourteenth edition, see _I.E.E. Publications Catalogue_, 1981.

42 _J.S.T.E.E._, XVII (May, 1888), 480.

43 _J.I.E.E._, XXVII (February, 1899), 212.

44 McKay, Tramways and Trolleys, p. 191.

NOTES FOR CHAPTER FIVE

1 Griffith, _Modern City Government_, p. 460.

2 Zeublin, _American Municipal Progress_, p. 43.

3 Dawson, _Life and Government in Germany_, p. 199.

4 _Ibid._, p. 235.

5 Great Britain, _Tramway Act, 1870_, 33 & 34 Vict., ch. 78.

6 L.C.C. Archives, Minutes of Highways Committee, 19 May 1903, pp. 814-15.

7 McKay, _Tramways and Trolleys_, pp. 91-93.

8 Howe, _European Cities at Work_, p. 113.

9 Ernest S. Griffith, _A History of American City Government: The Conspicuous Failure, 1870-1900_ (New York: Praeger, 1977), p. 182.

10 _Electrical Review_, New York, XLV (August, 1904), 221, citing U.S. Consular Reports on German street railways.

11 Albert Shaw, _Municipal Government in Continental Europe_ (New York: Century, 1885), pp. 349-53.

12 Dawson, _Life and Government in Germany_, p. 197.

13 P. Brose, _Uber die Vewenbarheit der Dynmomaschinen_ in Siemens archives, folder 25/Li 15, vol. 2, p. 27. Henceforth this archival source will be cited as S.A.A. followed by the folder number. See also Harold P. Brown, "The New Instrument of Execution," _North American Review_, CXLIX (1889), 578-590.

14 _J.I.E.E._, XXXI (November, 1902), 226.

15 _The Electric Lighting Act, Minutes of Evidence_, ix.

16 L.C.C. Archives, CL/HIG/1/1, Printed Documents, "Board of Trade Regulations."

17 <u>Electrical Review</u>, XXXI (November, 1892), 533.

18 <u>Electrical World</u>, XX (December, 1889), 367.

19 L.C.C. Archives, CL/HIG/1/5, Letter of the Highways Committee to the Board of Trade, 1889.

20 Ferranti and Ince, <u>Life and Letters of Ferranti</u>, pp. 57-60.

21 Howe, <u>European Cities at Work</u>, p. 341.

22 Sunderland Borough Archives, Lighting Committee, vol. A22, section 1/1, Circular of Municipal Electrical Association. This source will be cited as SDB (Sunderland District Borough) followed by volume and section number.

23 L.C.C. Archives, Minutes of Highways Committee, 1899. minute no. 7, p. 11.

24 Great Britain, Parliament, <u>Parliamentary Papers</u>, 1898, vol. IX, "Report of the Joint Select Committee on Electrical Energy," p. 609.

25 Manchester City Archives, Papers of the Chamber of Commerce, Annual Meeting, 1901.

26 Ballin, <u>Organisation of Electricity Supply</u>, p. 54.

27 Merz and McLellan Papers, 1.1/1. 1898, Preliminary Report, p. 6.

28 <u>The Electrician</u>, LX (January, 1908), 426.

29 L.C.C. Archives, CL/HIG/1/73, Parliamentary Committee Printed Documents, "The Supply of Electrical Energy," 1907.

30 Data from Merz and McLellan Papers, 1. 1/6, 1904, no. 80 and from <u>Garcke's Manual of Electrical Undertakings</u>.

31 Dawson, <u>Municipal Government in Germany</u>, pp. 227-28.

32 William H. Dawson, <u>Industrial Germany</u>, (London: Collins, 1912), pp. 150-158.

33 <u>Ibid</u>., p. 162.

34 Arthur Riedler, <u>Emil Rathenau und das Werden der Grosswirtschaft</u> (Berlin: J. Springer, 1916), pp. 78-79

35 Manchester City Archives, Papers of the Chamber of Commerce, 1901.

36 T.J.E.R., XII (December, 1884), 471.

37 Francis Jehl, Menlo Park Reminiscences (Dearborn, Mich.: Edison Institute, 1941), II, 785.

38 Wainwright, History of the Philadelphia Company, chap. 2.

39 Charles A. Beard, American City Government (New York: Arno reprint, 1970), pp. 192-205. This book was first published in 1912 by the Century Co. of New York.

40 Electrical Review, New York, XL (March, 1902), 339, quoting J. Frank Morrison, President of the National Electric Light Association.

41 Forest MacDonald, Insull (Chicago: University of Chicago Press, 1962), p. 124.

42 Electrical Review, New York, XL (February, 1902), 232-235.

43 Ibid., XLV (July, 1904), 182

44 The Times, London, July 5, 1904, P. 4.

45 L.C.C. Archives, CL/HIG/1/74, "The L.C.C. Electric Power Bill Report."

46 University of Newcastle Special Collections, Mss. Misc. 81, Charles Merz Manuscript, 1936, p. 14.

47 Ibid., p. 14.

48 I.E.E. Archives, Naest 14, Merz Papers, R. Nelson, "Proceedings at the Unveilling of a Memorial Tablet," 1946, p. 18.

49 R.K. Webb, Modern England (New York: Harper and Row, 1968), p. 455.

50 Electrical Review, New York, XLVIII (January, 1906), 64.

NOTES FOR CHAPTER SIX

1 Byatt, British Electrical Industry, pp. 20-21.

2 Brian Bowers, R.E.B. Crompton: Pioneer Electrical Engineer (London: H.M.S.O., 1969), p. 14.

3 D.G. Tucker, Gisbert Kapp, 1852-1922 (Birmingham: University of Birmingham, 1973), p. 3.

4 Siemens Archives Memorandum, Schuckert, p. 6.

5 R.E. Crompton, Reminiscences (London: Constable, 1928), pp. 131-35.

6 Silvanus Thompson, Dynamo-Electric Machinery (New York: F.N. Spoon, 1886), p. 20.

7 James Clerk Maxwell, "A Dynamic Theory of the Electromagnetic Field," Philosophical Transactions of the Royal Society of London, CLV (1865), 459-513.

8 Aitken, Syntony and Spark, p. 22.

9 James Grieg, John Hopkinson: Electrical Engineer. (London: H.M.S.O., 1970), p. 28.

10 W. Slingo and A. Brooker, Electrical Engineering (London : Longman's, 1891), p. 34.

11 McLaren, Rise of the Electrical Industry, pp. 125-27.

12 Verity's Ltd., Catalogue, 1911, p. 17, in Northwestern Museum of Science and Industry Archives, Manchester.

13 George Forbes, "Some Electric Lighting Stations in Europe," J.I.E.E., XVIII (January, 1889), 192.

14 Alexander Siemens, "Some Recent Improvements in Electric Light Apparatus," J.S.T.E., IX (1880), 161.

15 Electrical Review, New York, XIX (January, 1892), 3, commented, "His achievements ... have made his name great among inventors."

16 Ferranti Archives, Samuel Insull, *Central Station Electricity Service* (Chicago: privately printed, 1950), p. 216.

17 McLaren, *Rise of the Electrical Industry*, p. 170.

18 *Electrical Review*, XXX (February, 1892), 260.

19 Ferranti Archives, L 7, Lesco Prospectus, 18 September 1888.

20 *Ibid.*, L7, Annual General Meeting of Lesco, 13 March 1891.

21 *Ibid.*, L 8, G.W. Partridge, "Early Experiences in the Electrical Supply Industry," p. 76.

22 *Electrical Review*, XXX (April, 1892), 506.

23 Riedler, *Emil Rathenau*, p. 72.

24 Nikola Tesla, *Lectures, Patents, Articles* (Beograd, Yugoslavia: Nikola Tesla Museum, 1956), pp. 198-199.

25 Hans Wuger, *75 Jahre Drehstrom*, (Switzerland: Technorama, 1966), p. 5.

26 John O'Neil, *The Prodigal Genius: The Life of Nikola Tesla* (New York: Washburn, 1944), p. 65.

27 Prout, *Westinghouse*, p. 72.

28 Passer, *Electrical Manufacturers*, p. 279.

29 *The Electrician*, XXIV (February, 1890), 396-397.

30 D.S.L. Cardwell, *The Organization of Science in England* (London: Heineman, 1957), p. 124.

31 Tucker, *Gisbert Kapp*, p. 6.

32 Arthur Shadwell, *Industrial Efficiency: A Comparative Study of Industrial Life in England, Germany and America* (London: Longman's Green, 1923), p. 632.

33 *Ibid.*, p. 640.

34 Gordon W. Roderick and Michael Stephens, *Scientific and Technical Education in Nineteenth Century England* (New York: Barnes and Noble, 1973), p. 16.

35 *Ibid.*, p. 20.

36 Michael Argles, From South Kensington to Robbins: An Account of English Technical and Scientific Education since 1851 (London: Longman's, 1964), chap. 3.

37 Electrical Trades Directory for 1893, Biographical Sketches, lxiv.

38 S.A.A. 25/Ln/860, correspondence from Siemens Brothers (London) to Siemens and Halske (Berlin), 1902.

39 John Dummelow, The Metropolitan Vickers Electrical Company, 1899-1949 (Manchester, 1949), p. 13.

40 Electrical Review, LVI (January, 1905), 135.

41 Engineering, XLIX (January, 1905), 135.

42 Byatt, British Electrical Industry, p. 102.

43 Insull, Central Station Service, p. 215.

44 "Forty Years of Progress, Dr. Merz's Contribution to Technical Advance," Electrical Review, CXXIX (November, 1940), 370.

45 Leonard Reich, "Industrial Research and the Pursuit of Corporate Security," Business History Review, LIV (1980), 504-505.

46 An excellent account of the work of organized research departments is Benjamin Garver Lamme, Electrical Engineer: A Biography (New York: G.P. Putnam's Sons, 1926).

47 For example, see Prentiss M. Brown, Alexander Dow (New York: Newcomen Society of America, 1951). p. 14.

48 Electrical Review, New York, XL (January, 1902), 54-56, 374-375.

49 Byatt, British Electrical Industry, p. 187.

50 75 Jahre AEG (Berlin: AEG, 1958), p. 10.

51 Quoted in Hannah, Electricity before Nationalisation, p. 92.

52 D. Morgan, "Central Station Steam Power Generation," Westinghouse Engineer, X (January, 1950), 7.

53 Insull, Central Station Service, p. 219.

54 J.I.E.E., LII (December, 1913), 144-145.

NOTES FOR CHAPTER SEVEN

1 Alfred D. Chandler, Strategy and Structure: Chapters in the History of Industrial Enterprise (Cambridge: M.I.T. Press, 1962), p. 28.

2 Mathew Josephson, Edison (New York: McGraw Hill, 1959), p. 181.

3 James Conniff, The Energy People: A History of P.S.E. & G (Newark: Public Service and Gas Company, 1978). p. 34.

4 Journal of Gas Lighting, XXXV (January, 1885), 86.

5 Jehl, Menlo Park Reminiscences, III, 1085-1086.

6 Woodberry, Beloved Scientist, pp. 198-199.

7 Data from Journal of Gas Lighting, Electrical World, Electrotechnische Zeitscrift, and Passer, Electrical Manufacturers.

8 Great Britain, Electric Lighting Act, 1882, 45 & 46 Vic. 1.

9 Byatt, British Electrical Industry, p. 24.

10 Drake and Gorham Ltd., Electric Lighting Plant (London: Drake and Gorham, 1903), p. 2.

11 Electrical World, VII (January, 1886), 2.

12 T.J.E.R., XVI (October, 1885), 356.

13 Swale, Forerunners of the N.E.B., p., 26, quotes one observer in 1891 who commented "I have visited every electrical installation ... in England. Each installation has confirmed my early conviction that electric lighting will not pay."

14 Electrotechnische Anzeiger, I-II (October, 1884), 17.

15 Proceedings of the National Electric Light Association, Second Annual Convention, Baltimore, 1886, pp. 38-39.

16 Crompton, Reminiscences, p. 147.

17 T.J.E.R., XVII (July, 1885), 12; Brighton Corporation Electrical Department, Historical Notes (Brighton: BCED, 1903), pp. 1-4.

18 T.J.E.R, XVIII (January, 1886), 36.

19 Electrical World, III (November, 1884), 168.

20 Crompton, Reminiscences, p. 136; Ferranti and Ince, Life of Ferranti, p. 127.

21 J.I.E.E., XXX (February, 1901), 520-528.

22 Crompton, Reminiscences, pp. 131-135.

23 Electrical World, XV (February, 1895), 274-276. Electrical Engineer, XI (May, 1894), 404-405.

24 J.I.E.E., XXIV (June, 1892), 293-294.

25 The Statist estimated that between 1878 and 1883, 91 private companies were formed. After 1883 only about 8 solvent companies remained.

26 Jehl, Menlo Park Reminiscences, III, 9652.

27 Josephson, Edison, p. 270.

28 S.A.A. 46/Lc 739, Siemens Wirtschaftliche Mitteilungen; "Das Werden der Siemens Betriebe," 034.

29 Schukert, pp. 17-18, 40-41.

30 75 Jahre AEG, p. 9.

31 Chandler, Visible Hand, p. 194.

32 John Winthrop Hammond, Men and Volts; The Story of General Electric (New York: J.B. Lippencott, 1941), p. 89.

33 Electrical World, XIX (March, 1892), 177.

34 Ibid., XXXVII (January, 1902), 161.

35 Josephson, Edison, pp. 351-353.

36 Hammond, Men and Volts, pp. 220-224.

37 _Electrical Review_, XLVIII (May, 1901), 348; _Electrotechnische Zeitschrift_, XXII (January, 1902), 663, citing German company reports noting that foreign contracts were keeping workshops busy during the slump in the home market.

38 _Western Electrician_, XXX (January, 1902), 14-15; _Electrotechnische Zeitschrift_, XXII (January, 1902), 664.

39 _Schuckert_, pp. 66-70.

40 _75 Jahre AEG_, p. 29.

41 _Electrical Review_, XLVI (March, 1900), 384. A similar situation occurred in the engineering industry which also suffered from increased American imports; see S.J. Nicholas, "The American Export Invasion, The Case of the Engineering Industry, 1870-1914," _Technology and Culture_, XXI (October, 1980), 870-888.

42 _Electrical Review_, LVI (January, 1905), 1. This account of the English electrical industry noted that order books were full in 1900 in spite of the foreign competition.

NOTES FOR CHAPTER EIGHT

1 Passer, _Electrical Manufacturers_, p. 181.

2 Meeting of National Electric Light Association reported in _Electrical World_, XXVII (May, 1902), 931.

3 The major problem of central station managers in London was meeting the peak evening demand which was normaly about three times greater than the morning peak. L.C.C. archives, CL/HIG/1, Printed Documents, "The Electricity Supply of London," 1912, p. 2.

4 "Technical Aspects of the Period," in United States Bureau of the Census, _Central Electric Light and Power Stations_, (Washington: Government Printing Office,1910), pp. 120-122.

5 _J.I.E.E._, XXIV (January, 1895), 13.

6 Merz and McLellan Papers, 1/1.6, 1904.

7 _The Electrician_, XLII (January, 1899), 371.

8 Chandler, _Strategy and Structure_, p. 28.

9 Hammond, _Men and Volts_, pp. 193-195.

10 Merz and McLellan Papers, 1/1.1, no. 5; 1/1.5, 1903, no. 23, "Report on the Organization of a Consumer's Department."

11 _75 Jahre AEG_, p. 32.

12 _The Electrician_, XL (April, 1898), 859. The supply undertakings feared that the different voltage demands of motors would cause fuses and cut-outs to blow out.

13 Passer, _Electrical Manufacturers_, p. 264.

14 _Electrical Review_, XLVIII (February, 1901), 842.

15 _Electrical Industries and Investments_, VI (March, 1906), 649.

16 Comment of Herr Stettiner of the German General Electric Company, Journal of the Iron and Steel Institute, LXII (1902), 266.

17 Electrical Review, XLVII (August, 1900), 209

18 W. Paul Strassman, Risk and Technological Innovation: Manufacturing Methods in the Nineteenth Century (Ithaca, N.Y.: Cornell University Press, 1959), chap. 4.

19 Reports of the factory inspectors cited in Sidney Pollard and Paul Robertson, The British Shipbuilding Industry, 1870-1914 (Cambridge, Mas.: Harvard University Press, 1979), p. 121.

20 Merz and McLellan Papers, 1/1.11. 1908, "Report on the Supply to Harton Pits," p. 5.

21 Electrical Review, LIX (December, 1906), 974-975.

22 Merz and McLellan Papers, 1/1.11, 1909, "Report on the Development of Undertakings," 1/1.5, 1903, "Report on Proposed Electric Power Schemes."

23 Survey of electricity in mines carried out by Scott and Mountain Ltd., in Electrical Review, LVI (May, 1905), 779-780, summarizing several previous surveys.

24 From Telegraphs to Nuclear Power: Electrical Engineering in Cleveland and South Durham (Tees-side: I.E.E. Press, 1969), pp. 23-26.

25 W.O. Henderson, The Rise of German Industrial Power, 1834-1914 (Berkeley, Calif.: University of California Press, 1975), p. 178.

26 Great Britain, Parliamentary Papers, c.d. 2600, vol. XIV (1926), Report of the Samuel Commission.

27 Merz and McLellan Papers, 1/1.7, 1905, "Report on the Prospect of Power Supply in the Company's Area," November 1905, pp. 8-24.

28 The Engineer, XCIV (January, 1905), 7.

29 Colliery Guardian, XCV (March, 1908), 455.

30 Iron and Coal Trade Review, LX (December, 1901), supplement, 41.

31 Colliery Guardian, LXXXVII (February, 1904), 288.

32 Llewelyn Atkinson, "The Distribution of Power in Collieries," read at the S. Wales Institute of engineers and reported in the Electrical Review, XXXVII (October, 1895), 437-438.

33 Colliery Guardian, LXXXI (March, 1901), 695.

34 Journal of the Iron and Steel Institute, XLV (1894), 252; LXII (1902), 246.

35 "Electric Cranes and Capstan Equipment," Pamphlet, S.A.A. 25/Ln 862.

36 The Electrician Industrial Supplement, LIX (1901), 32.

37 Journal of the Iron and Steel Institute, LIX (1901), 32.

38 The Engineer, CII (July, 1906), 93.

39 Presidential Address of E. Windsor Richards, Journal of the Iron and Steel Institute, XLV (1894), 17-19.

40 John H. Carr and William Taplin, A History of the British Steel Industry (Cambridge, Mass.: Harvard University Press, 1962), p. 267.

41 The Electrician Industrial Supplement, LIX (July, 1907), 117.

42 Quoted in A.E. Musson, The Growth of British Industry (New York: Holmes and Meier, 1978), p. 194.

43 Merz and McLellan Papers, 1/1.8A, 1906, "Preliminary Report on Lancashire Electric Power," p. 1.

44 Byatt, British Electrical Industry, pp. 74-76.

NOTES FOR CHAPTER NINE

1 Electrotechnische Zeitschrift, XVV (April, 1914), 908-909.

2 Merz and McLellan Papers, 1/1.6, 1904, "Horsepower of Electricity Connected for Power and Light."

3 Engineering, XCVIII (December, 1904), 575.

4 Barker and Robbins, History of London Transport, II, 113-117.

5 Electrical World, XXXIII (March, 1901), 394-395.

6 Kenneth Hudson, Food, Clothes, and Shelter: Industrial Archeology in the Twentieth Century (London: John Baker, 1978), p. 27.

7 Electrical Review, LXXIV (April, 1914), 637.

8 Charles Edgar, "Electrical Progress of the Year," in Electrical Review, New York, XLIV (January, 1904), 41-42.

9 Proceedings of the Second Annual Conference on Education and the Training of Engineers reported Electrical Engineer, LII (june, 1911), 735.

10 J.I.E.E., XLVI (November, 1911), 131.

11 Quoted in Levine, Industrial Retardation in Britain, p. 75.

12 Speech of A.A. Day, J.I.E.E., L (November, 1912), 41-49.

13 Electrical Review, XLVI (March, 1900), 552-554.

14 Mellanby, History of Wiring, p. 43.

15 B Mathews, "Notes on the Commercial Development of Electricity Supply Undertakings," Electrical Review, LIX (October, 1906), 652-671, 728-729.

16 _The Electrician_, LIX (July, 1907), 491.

17 Figures from the German Association of Electrical Engineers, reprinted in _Electrical Review_, LXXIV (April, 1914), 637. _The Electrician_, LXVIII (February, 1912), 679, noted that the majority of the new German central stations were serving areas that did not have a supply of gas.

18 _The Electrician_, LX (December, 1907), 389.

19 McKay, _Tramways and Trolleys_, p. 66.

20 Hennessey, _Electrical Revolution_, pp. 36-39; Self and Watson, _Electricity Supply in Great Britain_, pp. 28-29.

21 Report of the Weir Committee appointed to review the National Problem of the Supply of Electrical Energy, cited in Hannah, _Electricity before Nationalization_, p. 91.

22 Ensor, _England, 1870-1914_, p. 508; average prices for electricity given in L.C.C. archives, CL/HIG/1/73, evidence before Lord Camperdown's Committee.

23 Hannah, _Electricity before Nationalization_, p. 61.

24 Self and Watson, _Electricity Supply in Great Britain_, p. 32.

25 Speech of S.D. Schofield, _J.I.E.E._, L (November, 1912), 50.

26 "The Electrical Engineer's Enquiry into the Electrical Industry in Great Britain," _Electrical Engineer_, LII (May-June, 1911), 551, 689-692, 721.

27 See A. Curtiss-Haywood, "Electricity and the Law," _The Electrician_, XXIV (February, 1889), 336.

BIBLIOGRAPHY

BOOKS AND PAPERS PUBLISHED BEFORE 1915

Beard, Charles A. American City Government. New York: Century Co., 1912

Bouver, George Spencer, and Webb Walter. The Law Relating to Electric Lighting. Sampson Law, Marston, Searle and Rivington, 1882.

Brighton Corporation Electrical Department. Historical Notes. Brighton: BCED., 1903.

Dawson, William H. Industrial Germany. London: Collins, 1912.

________. Life and Government in Germany. London: Longman's and Greene, 1914.

Deutschen Edison Gesellschaft. Das Edison Gluhlicht. Berlin: J. Springer, 1884.

Drake and Gorham Ltd. Electric Lighting Plant. London: Drake and Gorham. 1903.

Edison Electric Light Companies. Bulletins. 1882-1886.

Electric Lighting Act: Minutes of Evidence. London: Scientific Publishing Co., 1882.

The Ferranti System of Electric Lighting. London: Whitehead, Morris and Lowe, 1882.

Fortschritt der Electrotechnik 1886-1892. Berlin: J. Springer, 1887-1893.

Goodnow, Frank. The Modern Development of City Government. New York: Century Co., 1904.

Howe, Frederick C. European Cities at Work. New York: Charles Scribner's Sons, 1914.

Maxwell, James Clerk. "A Dynamic Theory of the Electromagnetic Field." Philosophical Transactions of the Royal Society, CLV (1865).

Oersted, Hans C. The Discovery of Electromagnetism made in the Year 1820. Copenhagaen: Larsen reprint 1920. First published in 1820.

Prescott, George. Dynamo Electricity. New York: Appleton, 1888.

Proceedings of the National Electric Light Association. Second Annual Convention. Baltimore: Baltimore Publishing Co., 1886.

Shaw, Albert. Municipal Government in Continental Europe. New York: Century Co., 1895.

Siemens and Halske. Electrische Bahnen. Berlin: Siemens and Halske, 1896.

Slingo, William, and Brooker, A. Electrical Engineering. London: Longman's, 1891.

Thompson, Silvanus. Dynamo Electric Machinery. New York: F.N. Spoon, 1886.

Verity's Ltd. Catalogue. 1911.

GOVERNMENT DOCUMENTS

Great Britain. _Electric Lighting Act_, 1882. 45 & 46 Vic. 1, ch. 56.

Great Britain. Parliament. _Hansard's Parliamentary Debates_. 3rd & 4th ser.

Great Britain. Parliament. _Parliamentary Papers_. Annual Reports of the Chief Inspector of Factories and Workshops, 1896-1900.

________. 1886. Vol. XXIII. C 4893. "Final report of the Royal Commission Appointed to inquire into the Depression of Trade and Industry."

________. 1889. Vol. LXX. C 189. "Report of the Marindin Commission."

________. 1898. Vol. IX (213). "Report of the Joint Select Committee on Electrical Energy."

________. 1910. Vol. CIX. Cd. 5005. "Final Report of the Census of Production."

________. 1926. Vol. XIV. Cd. 2600. "Report of the Samuel Commission."

Great Britain. _Tramways Act_, 1870. 33 & 34 Vic. 1, ch. 78.

United States. Bureau of the Census. _Central Electric Light and Power Stations_. Washington: Government Printing Office, 1910.

United States. Department of Commerce. _Historical Statistics of the United States_. Bicentennial Edition. Washington: Government Printing Office, 1975.

NEWSPAPERS AND PERIODICALS PRINTED BEFORE 1914

Collier's Magazine.

Colliery Guardian.

The Economist.

Electrical Engineer.

Electrical Industries and Investments.

Electrical Review. London.

Electrical Review. New York.

Electrical Trades Directory.

Electrical World.

The Electrician.

The Electrician Industrial Supplement.

Electrotechnische Zeitschrift.

The Engineer.

Engineering.

Garcke's Manual of Electrical Undertakings.

Iron and Coal Trades Review.

Journal of Gas Lighting.

Journal of the Iron and Steel Institute.

Journal of the Institution of Electrical Engineers.

Journal of the Society of Telegraph Engineers and Electricians.

Nineteenth Century.

Pall Mall Gazette.

The Statist.

Telegraph Journal and Electrical Review.

The Times.

Western Electrician.

BOOKS AND ARTICLES PRINTED AFTER 1914

Aitkin, Hugh G.J. Syntony and Spark: The Origins of Radio. New York: Wiley, 1976.

Appleyard, Rollo. The History of the Institution of Electrical Engineers. London: Institution of Electrical Engineers, 1939.

Argles, Michael. From South Kensington to Robbins: An Account of English Technical and Scientific Education since 1851. London: Longman's, 1964.

Ashworth, William. A Short History of the International Economy since 1850. London: Longman's, 1962.

Ausubel, Herman. The Late Victorians. New York: Van Nostrand, 1955.

Balin, H.H. The Organisation of Electricity Supply in Great Britain. London: Electrical Press, 1946.

Barker, T.C., and Robbins, Michael. A History of London Transport. Vol. II: The Twentieth Century. London: Allen and Unwin, 1974.

Bennett, Charles A. A History of Manual and Industrial Education. Peoria, Ill.: Manual Arts Press, 1937.

Bowers, Brian. R.E.B. Crompton: Pioneer Electrical Engineer. London: H.M.S.O., 1969.

________. "Electric Traction: Early Electric Motors." Institution of Electrical Engineers, 2nd Historical Weekend, July 1974, London.

Braun, Hans-Joachum. "Old and New Technologies: The Competition between Gas Lighting and Electric Lighting in Germany, 1880-1914." Unpublished Manuscript, Universitat Bochum, 1979.

Brown, Prentiss M. Alexander Dow. New York: Newcomen Society of America. 1951.

Byatt, Ian C.R. The British Electrical Industry, 1875-1914. Oxford; Clarendon Press, 1979.

Cardwell, Donald S.L. The Organisation of Science in England. London: Heinemann, 1957.

________. Technology, Science and History. London: Heineman, 1972.

Carr, John C., and Taplin, William. A History of the British Steel Industry. Cambridge, Mass.: Harvard University Press, 1962.

Chandler, Alfred D., Jr. Strategy and Structure: Chapters in the History of Industrial Enterprise (Cambridge, Mass.: M.I.T. Press, 1962.

________.The Visible Hand: The Managerial Revolution in American Business. Cambridge, Mass.: Harvard University Press, 1978.

A Chronological History of Electrical Development. New York: National Electrical Manufacturers Association, 1946.

Clapham, John H. An Economic History of Modern Britain. Vol. III: Machines and National Rivalries, 1887-1914. Cambridge: University Press, 1938.

Conniff, James. The Energy People: A History of PSE & G. Newark, N.J.: Public Service and Gas Company, 1978.

Crompton, R.E.B. Reminiscences. London: Constable, 1928.

Davies, Stephen. The Diffusion of Process Innovations. Cambridge: University Press, 1979.

Dummelow, John. The Metropolitan Vickers Company, 1899-1949. Manchester, 1949.

Dunsheath, Percy. A History of Electrical Power Engineering. Cambridge, Mass.: M.I.T. Press, 1962.

Ensor, Robert C.K. England, 1880-1914. Oxford: Clarendon Press, 1960.

Escher-Wyss, 1805-1955: 150 Years of Development. Zurich: Escher-Wyss, 1955.

Ferranti, Gertrude, and Ince Richard. The Life and Letters of Sebastian de Ferranti. London: Norgate and Williams, 1934.

Fox, Kenneth. _Better City Government: Innovation in American Urban Politics_. Philadelphia: Temple University Press, 1977.

Fraser, Derek. _Power and Authority in the Victorian City_. London: St. Martin's Press, 1979.

From Telegraphs to Nuclear Power: Electrical Engineering in Cleveland and South Durham. Tees-side: Institution of Electrical Engineers, 1969.

Garvin, John H. _The Life of Joseph Chamberlain_. London: MacMillan, 1932.

Goldman, Jacob E. "Science for Economic Growth and Social Change" _The Interaction of Science and Technology_. Edited by W. Dale Crompton. Urbana, Ill.: University of Illinois Press, 1969.

Griffith, Ernest S. _The Modern Development of City Government_. London: Oxford University Press, 1927.

________. _A History of American City Government: The Conspicuous Failure, 1870-1900_. New York: Praeger, 1977.

Greig, James. _John Hopkinson: Electrical Engineer_. London: H.M.S.O., 1970.

Griliches, Zvi. "Hybrid Corn: An Exploration of the Economics of Technological Change." _Econometrica_, XXV (October, 1957), 501-522.

Insull, Samuel. _Central Station Electricity Service_. Chicago: Privately Printed, 1950.

Hammond, John W. _Men and Volts: The Story of General Electric_. New York: J.B. Lippincott, 1941.

Hannah, Leslie. _Electricity before Nationalisation_. Baltimore: Johns Hopkins Press, 1979.

Henderson, William O. _The Rise of German Industrial Power, 1834-1914_ Berkeley, Calif.: University of California Press, 1975.

Hennessey, Richard A. _The Electrical Revolution_. Newcastle-on Tyne: Oriel Press, 1971.

Hudson, Kenneth. _Food, Clothes and Shelter: Industrial Archeology in the Twentieth Century._ London: John Baker, 1978.

Hughes, Thomas Parke. "British Electrical Industry Lag,. 1882-1888." _Technology and Culture_, III (1962), 27-44.

Hutchinson, Keith. _The Decline and Fall of British Capitalism_. New York: Charles Scribner's Sons, 1950.

Jehl, Francis. _Menlo Park Reminiscences_. Dearborn, Mich.: Edison Institute, 1941.

Josephson, Mathew. _Edison_. New York: McGraw Hill, 1959.

Landes, David. _The Unbound Prometheus_. Cambridge: University Press, 1869.

Lamme, Benjamin G. _Electrical Engineer: An Autobiography_. New York: G.P. Putnam's Sons.

Levine, Aaron L. _Industrial Retardation in England, 1880-1914_. London: Weidenfeld and Nicolson, 1972.

MacDonald, Forest. _Insull_. Chicago: University of Chicago Press, 1926.

McKay, John P. _Tramways and Trolleys: The Rise of Urban Mass Transit in Europe_. Princeton: Princeton University Press, 1976.

Maclaren, Malcolm. _The Rise of the Electrical Industry in the Nineteenth Century_. Princeton: Princeton University Press, 1943.

Mansfield, Edwin. "Technical Change and the Rate of Imitation." _Ecometrica_, XXIX (October, 1961), 741-66.

Martin, Thomas C. _Forty Years of Edison Service_. New York Edison Co., 1922.

Mathias, Peter. _The First Industrial Nation_. New York: Charles Scribner's Sons, 1969.

Maxwell, Bertram C. _Contemporary Municipal Government of Germany_. Baltimore: Warwick and York, 1928.

Mellanby, John. _The History of Electric Wiring_. London: MacDonald, 1957.

Mitchell, B.R. and Deane, Phyllis. _Abstract of British Historical Statistics_. Cambridge: University Press, 1962.

Morgan, D. "Central Station Steam Power Generation," Westinghouse Engineer, X (January, 1950), 7-20.

Mottelay, P. F. Biographical History of Electricity and Magnetism. London: C. Griffin, 1926.

Munro, William B. The Government of Anmerican Cities. New York: MacMillan, 1920.

Murphy, Brian. A History of the British Economy. London: Longman's, 1973.

Musson, Albert E. The Growth of British Industry. New York: Holmes and Meier, 1978.

Nicholas, S.J. "The American Export Invasion of Britain: The Case of the Engineering Industry, 1870-1914." Technology and Culture, IV (1980), 570-588.

O'Neil, John. The Prodigal Genius: The Life of Nikola Tesla. New York: Washburn, 1944.

Parker, Michael and Reid, D.J., The British Revolution. London: Blandford Press, 1972.

Passer, Harold C. The Electrical Manufacturers, 1875-1900. Cambridge: Harvard University Press, 1953.

Pollard, Sidney and Robertson, Paul. The British Shipbuilding Industry, 1870-1914. Cambridge: Harvard University Press, 1979.

Prout, Henry G. The Life of George Westinghouse. New York: Arno Press, 1972.

Redlich, Joseph and Hirst, Francis W. The History of Local Government in England. London: MacMillan, 1970.

Reich, Leonard. "Industrial Research and the Pursuit of Corporate Security." Business History Review, LIV (1980), 504-529.

Rice, Wallace. Seventy-Five Years of Gas Service in Chicago. Chicago: People's Gas Company, 1925.

Riedler, Arthur. Emil Rathenau und das Werden der Grosswirtschaft. Berlin: J. Springer, 1916.

Roderick, Gordon, W. and Stephens, Michael D. Scientific and Technical Education in Nineteenth Century England. New York: Barnes and Noble, 1972.

Rostow, Walt W. The World Economy: History and Prospect. Austin, Tex.: University of Texas Press, 1978.

Rowland, John. Progress in Power. London: Newman Neame, 1966.

Sahal, Devandra, ed. Research, Development and Technological Innovation. Lexington, Mass.: D.C. Heath, 1980.

Sakamoto, Takuji. "Introduction and Development of Technology in the British Electrical Machinery Industry, 1875-1914." Paper given at the First International Conference of Business History, Shizuoka, Japan, 1979.

Saul, S.B. The Myth of the Great Depression, 1873-1896. London: MacMillan, 1969.

Self, Sir Henry, and Watson, Elizabeth. Electricity Supply in Great Britain. London: George Allen and Unwin, 1952.

75 Jahre AEG. Berlin: AEG, 1969.

Shadwell, Arthur. Industrial Efficiency: A Comparative Study of Industrial Life in England, Germany and America. London: Longman's Green, 1923.

Sharlin, Harold. The Making of the Electrical Age. London: Abelard Schuman, 1963.

Silverberg, Richard. Light for the World: Edison and the Power Industry. Princeton: Van Nostrand, 1967.

Singer, Charles et al. A History of Technology. New York: Oxford University Press, 1958.

Sprague, Harriet. Frank J. Sprague and the Edison Myth. New York: William Frederick, 1947.

Stotz, Louis. The History of the Gas Industry. New York: Stettiner, 1938.

Strasseman, W. Paul. Risk and Tchnological Innovation: American Manufacturing Methods in the Nineteenth Century. Ithaca, N.Y.: Cornell University Press, 1959.

Swale, R.E. Forerunners of the Northeastern Electricity Board. Manchester; Northeastern Electricity Board, 1963.

Tames, Richard. Economy and Society in Nineteenth Century Britain. London: George Allen and Unwin, 1972.

Tesla, Nikola. Lectures, Patents, Articles. Boegrad, Yugoslavia:

Nikola Tesla Museum, 1956.

Tucker, D.G. Gisbert Kapp, 1852-1922. Birmingham: University of Birmingham, 1973.

________. "W.H. Preece, Nineteenth Century Telegraph, Telephone and Power Station Engineer." I.E.E. Historical Paper, 1974.

Wainwright, Nicolas. History of the Philadelphia Electric Company. Philadelphia: Philadelphia Electric Co., 1961.

Warner, Sam Bass. Streetcar Suburbs: The Process of Growth in Boston, 1970-1914. Cambridge: Harvard University Press, 1978.

Webb, Robert K. Modern England: From the Eighteenth Century to the Present. New York: Harper and Row, 1968.

Weiher, Siegfried von, and Goetzler, Herbert. Weg und Werken der Siemenswerke in Forthschritt der Elektrotechnik. Munich: Bruckman, 1972.

Williams, Leslie Pearce. Michael Faraday: A Biography. New York: Basic Books, 1965.

Woodbury, David, Beloved Scientist: Elihu Thompson. New York: McGraw Hill, 1944.

Wuger, Hans. 75 Jahre Drehstrom. Switzerland: Technorama, 1966.

Young, G.M. Portrait of an Age. London: Oxford University Press, 1969.

Zeublin, Charles. American Municipal Progress. New York: MacMillan, 1916.

ARCHIVAL SOURCES

Institution of Electrical Engineers, London.
Some of the leading electrical engineers left their papers to the Institution, including Charles Brush. The library contains a fine collection of nineteenth century books and journals.

Tyne and Wear Archives, Newcastle.
Contains the papers of the Merz and McLellan consulting firm.

Northeastern Electricity Board, Newcastle.
The main office of this organization contains printed material related to the early years of electrical supply in the Northeast, and many papers of Nesco.

Sunderland Borough Archives, Sunderland.
Has documents of the Municipal Electrical Association.

Ferranti Collection, Hollingsworth near Manchester.
The largest and most complete source of company records for the period 1880-1914.

Northwestern Museum of Science and Industry, Manchester.
Contains many pamphlets and printed material of the electrical industry in the nineteenth century.

Manchester City Archives, Manchester.
Contains records of the Chamber of Commerce and other business organizations in Manchester.

London County Council (Now GLC), London.
Houses all records of this important municipal government, including parliamentary documents.

Milne Museum of Electrical History, Tonbridge, Kent.
This museum has a fine collection of artifacts as well as printed material from the 1880s and 1890s.

Siemens Archives, Munich.
Contains copious records of the activities of the Siemens and Halske company which cover its worldwide activities. The Siemens Archives library had an unsurpassed collection of books related to the history of electricity.

General Electric Archives, Schenectady, New York.
This archive has little unpublished material on the early years of the company, but it does contain documentation on the meetings of the National Electric Light Association.

Edison National Historic Site, West Orange, New jersey.
Edison's original laboratory houses millions of pages of documents generated by Edison and his many electrical companies. These are currently being edited and will be published over the next twenty years.

INDEX